The Development of Darwin's Theory

Charles Darwin. (From R. E. Keynes, ed., *The Beagle Record*, Cambridge University Press, 1979, p. 14.)

THE DEVELOPMENT OF DARWIN'S THEORY

Natural History, Natural Theology, and
Natural Selection, 1838–1859

DOV OSPOVAT
Late Associate Professor of History
University of Nebraska – Lincoln

CAMBRIDGE
UNIVERSITY PRESS

Published by the Press Syndicate of the University of Cambridge
The Pitt Building, Trumpington Street, Cambridge CB2 1RP
40 West 20th Street, New York, NY 10011-4211, USA
10 Stamford Road, Oakleigh, Melbourne 3166, Australia

First published 1981
First paperback edition 1995

Library of Congress Cataloging-in-Publication Data
Ospovat, Dov.
The development of Darwin's theory.
Bibliography: p.
1. Darwin, Charles, 1809–1882.   2. Natural selection –
History.   3. Natural history – History.   4. Biology –
History.   5. Naturalists – England – Biography.
I. Title.
QH31.D2074   575.01′62   81–4077
AACR2

ISBN 0 521 46940 6 Paperback

Transferred to digital printing 1999

To Claudia, Conoley, Sam,
Joyce, and Sascha

Contents

Illustrations

Foreword

On a hot summer's day in 1977, Dov Ospovat and I drove through South London. Dov had never seen Richard Owen's 30-ton, concrete dinosaur restorations, standing today exactly where Owen had left them in 1854 in the grounds of the gutted Crystal Palace. As we drove in the decayed splendour of my old Jaguar across the Thames we chatted about the rehabilitation of Richard Owen.

Dov was working on Owen, arguably England's greatest comparative anatomist. Owen was still a hate figure, the zoological autocrat, and was duplicitous, or so Darwin had supposed. With the Darwinian "victors" writing history, Owen had never been given a good press. But he was a superb, if plodding, morphologist and paleontologist, and Dov's was the first sympathetic reappraisal of his science. In 1977 Dov had started to publish good, suggestive papers, stemming from his 1974 Harvard Ph.D. He showed how Owen in the 1850s – before Darwin had begun to write the *Origin of Species* – was modifying von Baer's embryology to produce a ramifying, specializing, implicitly tree-like picture of fossil succession.[1]

This image, Dov realized, was Darwin's too. It began to look as though Owen had played an unwitting John the Baptist to Darwin's Christ. Dov was already thinking of Darwin as he photographed Owen's rhinoceros-like *Iquanodon* at Crystal Palace. He wanted to see Darwin's work related to the dominant biology of the 1830s–1850s, to Owen's biology. He wanted to make Darwin less of a seer, standing out of time, and more a man of his day.

That was the last time I saw Dov. He went back to Lincoln, Nebraska, where he taught history of science. A string of papers ensued, and he skewed more and more into Darwin studies. (I remonstrated with him jokingly, but we all go that way eventually!) And once onto Darwin, he made surprising headway. He showed that, after Darwin devised the theory of natural selection in 1838, he still believed that organisms remained perfectly adapted, at least until geological conditions changed. Only then did a struggle and selection operate to keep the species perfectly fitted to its changing niche. Darwin continued to inhabit Paley's – and Owen's – natural theological world. No scholar before Dov had realized this before. Dov reckoned that Darwin only jettisoned the notion of perfection and shifted to a belief in relative adaptation in 1854. And this affected Darwin's whole evolutionary thinking. Since chance always left some offspring better adapted than others, Darwin now saw that organisms were *permanently* engaged in an intra-specific struggle.[2] Natural selection operated continually. Dov had the makings of a book: Darwin's changing views in the two decades before 1859, when the *Origin of Species* was published.

I had a note from Dov in January 1980, saying that he had just posted off "a long manuscript call[ed] 'The Development of Darwin's Theory, 1838–59,'" a book that would sum up his reappraisal of Darwin in Owen's day. I was deep in my own *Archetypes and Ancestors* at the time, which would look at the ideological contexts of Owen's and Darwin's paleontology, and which was itself a social corollary of Dov's work. By 14 March 1980, Dov hadn't heard of his book's fate. "I doubt it will be out by Spring '81," he wrote.

I never heard from him again. He died of cancer on 28 September 1980. He was 33.

The Development of Darwin's Theory was his epitaph, the best any man could wish for. David Kohn helped to see it into print, and Cambridge University Press published it in 1981. Dov, I know, would have been quietly proud of it, and rightly. It was the first of the new-style books, and it puts Darwin squarely into the early Victorian period. There is no "presentism" here, no glancing back to see what gems of modern "truth" the perspicacious Darwin had collected. The study locates him firmly in young Victoria's day.

The book displayed all of Ospovat's ideas: on Darwin's relations with the philosophical anatomists of his day, on the archetypes and higher design, on the discovery of the "principle of divergence," and the shift away from the idea of perfect adaptation – in a nutshell, on Darwin's response to the "leading architects" of natural history. The book is also about Darwin's strategic presentation. He had to convince fellow experts that selection accounted for their scientific findings. As Dov says, much of Darwin's "effort from 1838 to 1859 was directed toward finding evolutionary explanations for generalizations his contemporaries proposed in their studies of morphology, embryology, paleontology, classification, and geographical distribution."

The years following the publication of the *Development of Darwin's Theory* saw a series of breakthroughs in surrounding areas. Darwin's notebooks were definitively transcribed and annotated.[3] The foremen in the Darwin Industry produced a number of scintillating analyses of Darwin's creative London years (1837–9), after the *Beagle* voyage.[4] In particular, Phillip Sloan exquisitely detailed Darwin's and Owen's converging views on active matter, life forces, and the duration of species, all of which were reflected in Darwin's 1837 speculations on transmutation.[5] No one had shown Darwin toying with Owen's ideas on matter and vitality at such a deep level before.

Dov set many hares running, and we all chased them. He pioneered a study of the philosophical anatomists: medically trained men such as Owen, W. B. Carpenter, Peter Mark Roget, and Martin Barry. They looked to a higher design based on the ground-plans of life, the great archetypes – and again Ospovat was pointing out how similar Darwin was to the progressive morphologists of his day. To Dov's list Philip Rehbock has added Robert Knox (of the Burke and Hare scandal). My interest was in another philosophical anatomist, Darwin's Edinburgh mentor Robert E. Grant, and I went on to delineate the social milieu of Dov's philosophical factions during the radical 1830s.[6]

But few have followed Dov's extraordinary furrow through the years leading up to 1859.[7] The *Development of Darwin's Theory* is still an arguing point. It was – and is – an immensely influential book, even if, as David Kohn argues, the roots of relative adaptation were present in Darwin's creative phase (1837–9).[8] The very existence of Darwin's two-decade delay has raised a fundamental question: Why did he refrain from publishing his theory of evolution for so long?[9] Was it simply that he had scientific business to finish, or did fears of ostracism stay the squire's hand?

In the 1990s, the great morphologist rehabilitated by Ospovat – Richard Owen – has at last come into his own. Phillip Sloan has transcribed Owen's 1837 Hunterian Lectures and provided an excellent commentary, and Kevin Padian has unearthed the manuscript of another of Owen's influential 1837 lectures. I have studied the politics of Owen's science in the unreformed College of Surgeons in London; Jacob Gruber and John Thackray are working on Owen's manuscript letters, and Nicolaas Rupke has produced a fine study of Owen's campaign for a natural history museum in London.[10]

And so fourteen years after Dov's death, his broader view prevails. We all take Richard Owen's philosophical anatomy seriously today, and we all see Darwin as a product of his time. Indeed, Robert Richards's claim that Darwin himself was locked into the transcendental movement caused a disbelieving Michael Ruse to title a review in 1993 "Were Owen and Darwin *Naturphilosophen?*"[11] So far has the pendulum swung Ospovat's way!

Were Dov alive today – and wandering again on a hot summer's day among Owen's dinosaurs – he would be well pleased with the way things have turned out.

ADRIAN DESMOND
Summer 1994

Notes

1 Dov Ospovat, "The Influence of Karl Ernst von Baer's Embryology, 1828– 1859: A Reappraisal in Light of Richard Owen's and William B. Carpenter's 'Palaeontological Application of "Von Baer's Law,"' " *Journal of the History of Biology*, 9 (1976): 1–28.
2 Dov Ospovat, "Perfect Adaptation and Teleological Explanation: Approaches to the Problem of the History of Life in the Mid-Nineteenth Century," *Studies in History of Biology*, 2 (1978): 33–56; "Darwin after Malthus," *Journal of the History of Biology*, 12 (1979): 211–230.
3 P. J. Gautrey, S. Herbert, D. Kohn, and S. Smith, eds., *Charles Darwin's Notebooks, 1836–1844*, Cambridge: British Museum (Natural History)/ Cambridge University Press, 1987.
4 Frank J. Sulloway, "Darwin and His Finches: The Evolution of a Legend," *Journal of the History of Biology*, 15 (1982): 1–53; "Darwin's Conversion: The *Beagle* Voyage and its Aftermath," *Journal of the History of Biology*, 15 (1982): 325–96. M. J. S. Hodge, "Darwin and the Laws of the Animate Part of the Terrestrial System (1835–7): On the Lyellian Origins of his Zoonomical Explanatory Program," *Studies in History of Biology*, 6 (1983): 1–106.
5 Phillip R. Sloan, "Darwin, Vital Matter, and the Transformation of Species," *Journal of the History of Biology*, 19 (1986): 369–445, esp. 399ff. The "vital materialism" of the German biologists who influenced Owen, and indirectly Darwin, has been described by Timothy Lenoir in *The Strategy of Life: Teleology and Mechanics in Nineteenth-Century German Biology*, Dordrecht: Reidel, 1982.

6 Philip F. Rehbock, *The Philosophical Naturalists: Themes in Early Nineteenth-Century British Biology*, Madison: University of Wisconsin Press, 1983. Evelleen Richards, "The 'Moral Anatomy' of Robert Knox: The Interplay between Biological and Social Thought in Victorian Scientific Naturalism," *Journal of the History of Biology*, 22 (1989): 373–436. Adrian Desmond, *The Politics of Evolution: Morphology, Medicine and Reform in Radical London*, Chicago: University of Chicago Press, 1989. Darwin's debt to Grant's invertebrate zoology is well brought out in Phillip R. Sloan, "Darwin's Invertebrate Program, 1826–1836: Preconditions for Transformism," in *The Darwinian Heritage*, ed. David Kohn, Princeton: Princeton University Press, 1985.
7 Although Janet Browne was simultaneously working on Darwin's "principle of divergence": Janet Browne, *The Secular Ark: Studies in the History of Biogeography*, New Haven: Yale University Press, 1983, pp. 210–16.
8 David Kohn, "Darwin's Ambiguity: The Secularization of Biological Meaning," *British Journal for the History of Science*, 22 (1989): 215–39. Peter J. Bowler, *Charles Darwin: The Man and His Influence*, Oxford: Blackwell, 1990, chap. 6.
9 Adrian Desmond and James Moore, *Darwin*, London: Michael Joseph, 1991. Robert J. Richards, *Darwin and the Emergence of Evolutionary Theories of Mind and Behavior*, Chicago; University of Chicago Press, 1987, pp. 142–52.
10 Phillip R. Sloan, ed., *Richard Owen: The Hunterian Lectures in Comparative Anatomy, May and June 1837*, Chicago; University of Chicago Press, 1992. Kevin Padian's transcription of a previously missing 1837 Hunterian Lecture by Owen, "Modifications of the Vertebral Column," is forthcoming in the *Journal of the History of Biology*. Desmond, *Politics of Evolution: Morphology, Medicine and Reform in Radical London*, Chicago: University of Chicago Press, 1989, chaps. 6–8. Jacob W. Gruber and John Thackray, *Richard Owen Commemoration*, London: Natural History Museum Publications, 1992. Nicolaas A. Rupke, *Richard Owen: Victorian Naturalist*, New Haven: Yale University Press, 1994. Evelleen Richards, "A Question of Property Rights: Richard Owen's Evolutionism Reassessed," *British Journal for the History of Science*, 20 (1987): 129–71.
11 Michael Ruse, "Were Owen and Darwin *Naturphilosophen*?" *Annals of Science*, 50 (1993): 383–8. Robert J. Richards, *The Meaning of Evolution: The Morphological Construction and Ideological Reconstruction of Darwin's Theory*, Chicago: University of Chicago Press, 1992.

Preface

I first read a few of Darwin's unpublished notes in the summer of 1974, at the suggestion of David Kohn and through the kindness of Sydney Smith. I was then working on a study of pre-Darwinian natural history, and I was curious to see how Darwin had reacted to the important changes that were occurring in the natural history sciences during the years when he was at work on his theory. David told me I would probably be interested in some of the items in the "Black Box" at Cambridge University Library. When I returned to Cambridge three years later for a longer look, I was struck by the vast quantity of notes, virtually ignored by Darwin scholars, that remained from the period 1838–59. The significance of this material became apparent to me in the fall of 1977 as I arranged my transcriptions of it in chronological order and examined it in conjunction with Darwin's transmutation notebooks and his two essays written in the 1840s. I knew from David's study of the transmutation notebooks and from my own recent work on Richard Owen, Louis Agassiz, and others that until he read Malthus Darwin believed in the natural theological idea of perfect adaptation, in the same sense as several of his contemporaries. I found now that Darwin continued to believe in perfect adaptation until the mid-1850s, which indicated that the theory of natural selection had evolved in previously unsuspected ways between Darwin's reading of Malthus and the publication of the *Origin of Species*. This suggested that it might be worthwhile to look more closely at the pre-*Origin* development of natural selection. I was the more willing to undertake such a task because, apart from numerous studies of the origins of the theory around 1838, its history before 1859 had not been the subject of any sustained investigation.

My greatest intellectual debts are to Camille Limoges and,

especially, David Kohn for their interpretations of the genesis of Darwin's theory. I have frequently disagreed with them, but without their insights, it is unlikely that my own study would ever have been conceived. Equally important was Sydney Smith's decision, after an acquaintance with me of only half an hour's duration, to introduce me to some of the contents of the Black Box and to share with me his ideas on their place in Darwin's work.

Many other people – teachers, librarians, colleagues, friends – have offered encouragement, advice, hospitality, and assistance of various kinds. I think particularly of Toby Appel, Mike and Jan Bartholomew, Janet Browne, Frederick Burkhardt, I. Bernard Cohen, William Coleman, E. H. Cornelius, Peter Gautrey, Gerry Geison, Stephen Gould, John Greene, Sandra Herbert, Ludi Jordanova and Karl Figlio, Terry Kohn, Malcolm and Dorian Kottler, Bari and Evelyn Logan, Nancy Mautner, Everett Mendelsohn, Jim Moore, Claudia Ospovat, Jeanne Pingree, Stan Rachootin, M. J. Rowlands, and Richard Ziemacki. I wish also to thank for their friendly assistance the staffs of the Library of the Royal College of Surgeons, the Library of the British Museum (Natural History), the Archives of Imperial College, London, the Manuscript Reading Room of Cambridge University Library, and the Life Sciences Library, Geology Library, and Love Memorial Library, all at the University of Nebraska – Lincoln.

For me there are many pleasant memories associated with the writing of this book. My only regret is that the satisfaction I have derived from it cannot be used to repay Claudia, Conoley, and Sam for what I took from them while I worked. Lost hours are lost forever.

D. O.

Lincoln, Nebraska
June 6, 1980

Acknowledgments

The first half of Chapter 1 is a revised version of an article that originally appeared in *Studies in History of Biology* (vol. 2, 1978, pp. 33–56), edited by William Coleman and Camille Limoges and published by The Johns Hopkins Press: "Perfect Adaptation and Teleological Explanation: Approaches to the Problem of the History of Life in the Mid-Nineteenth Century" (copyright 1978 by The Johns Hopkins Press). Chapter 3 is an expanded version of an article that first appeared in the *Journal of the History of Biology* (vol. 12, 1979, pp. 211–30), edited by Everett Mendelsohn, and is reprinted by permission of D. Reidel Publishing Company: "Darwin after Malthus" (copyright 1979 by D. Reidel Publishing Co., Dordrecht and Boston). I am grateful to both publishers for allowing me to use these articles. I acknowledge with thanks the permission granted by the Syndics of Cambridge University Library to quote from Darwin manuscripts in their possession; the permission granted by the Librarian of the British Museum (Natural History) to quote from the Richard Owen Collection; the permission granted by the Librarian of the Royal College of Surgeons to quote from the Owen manuscripts in its possession; and the permission granted by the Archives of Imperial College, London, to quote from materials in the Thomas Henry Huxley papers. The Research Council of the University of Nebraska – Lincoln awarded me a Maude Hammond Fling Faculty Research Fellowship and two research grants-in-aid, which made possible visits to Cambridge University Library in the summers of 1977 and 1978. I appreciate their generous support for this project.

Note on manuscript citations

At the time research on this book was done, a large quantity of Darwin's MS material was quite literally contained in a black trunk (known as the Black Box) in Cambridge University Library. The Black Box contained, among other manuscripts, a large collection of notes in Darwin's original portfolios. Although other scholars have examined these portfolios, perhaps no one has worked with them so intensely as the author. While this book was being written, the contents of the Black Box were removed and catalogued, the portfolios that the author studied under the heading DAR 205.

Dov Ospovat died on September 28, 1980, after several months' illness, before he could replace his old citations with the new numbers. Therefore, in the text and in comments in the Notes, references to the Black Box have been left intact. In the citations, the current catalog numbers have been used. These were furnished by David Kohn, to whom the Cambridge University Press and Dov Ospovat's family are extremely grateful.

Introduction: Darwin and his fellow naturalists

I have promised in the title of this book to discuss the development of the theory of natural selection between 1838 and 1859. This requires some explanation, because until now the most important phase in the "development of Darwin's theory" has generally been supposed to have occurred before the end of 1838, when Darwin's search for an adequate theory of transmutation was brought to a successful conclusion, or at least by 1844, when he wrote a pretty full exposition of his theory in a form that closely resembles that of the *Origin of Species*.[1] I argue here that between 1844 and 1859 the theory of natural selection was substantially transformed, in ways that are of interest for what they reveal about Darwin's ideas and about the intricate web of relationships between his creative scientific work and the biological thought of his contemporaries.

In July 1837, several months after returning from his five years' voyage on the *Beagle*, Darwin began keeping a series of notebooks on transmutation, in which during the next two years he recorded his speculations on the mode of production of new species.[2] In September 1838 he read Malthus's *Essay on the Principle of Population* and hit upon the idea of natural selection. The argument he constructed for his theory was complete, at least in outline, by the time he wrote his "Sketch of 1842." Two years later he expanded this into the long "Essay of 1844," which he thought was publishable if by chance he should die young. Why Darwin did not publish his theory at this time, and instead devoted eight years (1846–54) to morphological and taxonomical researches on barnacles, is a question that has generated much discussion.[3] Another question, perhaps more fruitful, has been virtually ignored: What was the result of his not publishing in 1844?[4] I do not go so far as to say that Darwin adopted a wholly new theory after 1844, but entertaining that possibility is a useful aid in gaining an apprecia-

1

tion of the magnitude of the changes that occurred in his ideas about evolution[5] and natural selection. Furthermore, it opens to investigation aspects of Darwin's thought and, more generally, a period in the history of biology which have usually been treated as of marginal importance for understanding the Darwinian revolution.

From September 1854, when the investigations on barnacles were concluded, until June 1858, when Wallace's letter arrived announcing his independent formulation of the theory of natural selection, Darwin worked full time on his species theory. In 1856 he began writing the big book, now published as *Natural Selection*, which he abandoned after the receipt of Wallace's letter in order to produce quickly an abstract of it – the *Origin of Species*.[6] During the period of intense work on his theory between 1854 and 1858, lines of thought that were initiated in the 1840s led Darwin to a new view of the evolutionary process. As late as 1844, the structure of Darwin's theory was to a large extent determined by what, for convenience, I call natural theological ideas or assumptions, assumptions that Darwin had held since before opening his first transmutation notebook. The transformation of the theory that occurred in the 1850s eliminated some of these assumptions, while others subsequently played a much less important role than they had previously. At the same time, it produced some of the most characteristically Darwinian ideas that we associate with the theory of natural selection. The idea of relative adaptation, for instance, is a product of the 1850s and is not to be found in the notebooks or the "Sketch" or "Essay."

The work of every scientist is necessarily dependent to a greater or lesser degree on ideas produced by his predecessors and contemporaries. This is one obvious way in which science is a social activity: The generation of scientific ideas is a historical process that involves the communication of ideas, beliefs, and assumptions. The natural theological ideas that informed Darwin's theory in its early years were prominent parts of an early-nineteenth-century British way of seeing nature, a way that was inculcated through the writings of theologians, scientists, scientific populariz-ers, and political economists. They included the idea – indeed the perception – that the adaptation of organisms to their environ-ments is perfect, that nature is a well-adjusted mechanism, that there is a harmony among organisms and between them and the inorganic world;[7] the idea that the laws of nature were established by God to achieve his ends; and the idea that all natural phenomena serve purposes relative to the whole economy of nature – for instance, that variation from the norm of the species is for the purpose of accommodation to new external conditions. These ideas, though not confined to Great Britain, were common in British natural theology and in the scientific writings on which Darwin depended most heavily for his education as a naturalist, such as Lyell's *Principles of Geology*. After 1859 Darwin's theory contributed to the complex process, already well under way, by which they lost currency. But for many years, and in some respects throughout his life, Darwin shared his contemporaries' belief in harmony and perfection. In constructing theories of transmutation in the period 1837–8, including the theory of natural selection, Darwin took for granted that adaptation is perfect and that variation is for the purpose of enabling organisms to accommodate to environmental change. Through these assumptions a view of nature common since the time of Robert Boyle, John Ray, and Isaac Newton contributed to and shaped Darwin's theories. Within a few months of reading Malthus, Darwin rejected most of natural theology's explicit formulations about nature and worked out his own ideas of design and purpose. But until the 1850s he adhered to assumptions that were deeply embedded in the traditional view, and these gave the theory of natural selection a particular structure, the structure of a mechanism of adjustment to change, a means by which the balance of nature is preserved. Since this feature of Darwin's early expositions of his theory has scarcely been noticed,[8] I discuss it at some length. Its construction and subsequent transformation are what I principally have in mind when I speak of the development of Darwin's theory.

On occasion, and it was true of Darwin, a scientist concludes that in order to present his new ideas for debate in the most effective manner it is necessary to undertake a sweeping reconstruction of existing knowledge. The necessity, it might be said, is imposed by the fact that the new ideas will be judged in part on their ability to account for the phenomena that contemporary science says are important. Here the social character of the scientific enterprise is particularly pronounced, for at every stage of the reconstruction, the new interpretations that are produced depend directly on the old interpretations that are being revised. The body of knowledge being reconstructed sets the problems the scientist must resolve in order to effect the transformation he has in view. And in the process, much of the old is inevitably incorporated, in more or less altered form, into the new. This occurred in Darwin's reconstruction of mid-nineteenth-century natural history, a project to which he committed himself in the transmutation notebooks. Much of his creative effort from 1838 to 1859 was directed toward finding evolutionary explanations for generalizations his contemporaries proposed in their studies of morphology, embryology, paleontology, classification, and geographical distribution. In setting problems for Darwin to resolve, his contemporaries unknowingly shaped the development of Darwin's theory, both in the details of his reconstruction of their science and in a more general and important sense: It was by way of his speculations on the problems their work posed that Darwin came to elaborate in the 1850s a new conception of the evolutionary process. In this new conception, the idea of perfect adaptation played no role; and instead of a mechanism of adjustment to external change, of preserving balance and harmony, natural selection became a theory in which the production of adaptation – which remained for Darwin the most important goal of the process – was held also to lead inevitably to the progressive development of life. In this way the socially imposed task of reinterpreting existing generalizations led to the mature theory of the *Origin of Species*. It caused Darwin to undertake pieces of work without which his published theory might have retained the natural theological structure it had in the "Essay of 1844."[9]

The twenty-one-year period between Darwin's reading of Malthus and the publication of the *Origin of Species* readily suggests the analogy of the growth of an individual from birth. It is an analogy that seems to me appropriate in one particular sense. In 1838 natural selection was mostly unrealized and not wholly determinate potential. If some of the characteristics it would have were inevitable consequences of Darwin's first insight on reading Malthus, many others were due to the circumstances in which it developed during the following years. By 1838 Darwin was a professional naturalist.[10] He had begun to immerse himself in the literature of natural history, and it was this literature that provided the most immediate circumstances in which the theory developed. So while the focus of this book is Darwin's theory, it is also about the theories of some of his contemporaries. Although no one would deny the importance in a study of Darwin's ideas of taking into consideration the ideas of his professional colleagues, it is surprising how seldom it is actually done. Rarely in Darwin studies are Karl Ernst von Baer, Henri Milne Edwards, Louis Agassiz, Richard Owen, and other leading architects of the theoretical natural history of Darwin's day accorded serious attention for their contributions to Darwin's thought or to the emergence of evolutionary biology after 1850.[11] I treat them here because of their importance to our understanding of the course of Darwin's speculations and of the evolutionary synthesis he produced. In emphasizing the desirability of studying Darwin's thought in the context of the thought of his fellow naturalists, I do not mean to imply that this context is more important than others, such as the political and economic changes and the political and economic ideas of the period.[12] It is one of several contexts that require attention before we can construct an adequate picture of the development of Darwin's theory and its impact on biologists and others after 1859. Because it was the most immediate context in which Darwin worked, consideration of it and Darwin's relationship to it may suggest some of the channels through which the more general social context indirectly exerted its influence on Darwin's science.

1

Darwin and the biology of the 1830s: some parallels

When he left England on H.M.S. *Beagle* in 1831, Darwin believed, with most of his contemporaries, that each species has been independently created with characteristics that suit it admirably for the conditions under which it is destined to live. By the spring of 1837 he was a transmutationist, believing that each species has descended from some other previously existing species and that its characteristics have been determined largely by heredity. This same brief interval also saw the beginnings of a profound alteration in biological thought generally – namely, the rejection by many of the best young British and continental naturalists of the teleological approach to biological explanation, which Georges Cuvier and British natural theology alike declared to be the only sure path to the understanding of organisms. The rejection of teleological explanation by Darwin's contemporaries provides the best perspective from which to examine one of the central elements in Darwin's transmutationist thought – his concept of adaptation. His acceptance of transmutation immediately led Darwin too to reject teleological explanations of biological phenomena, and many of his arguments against "creation" in both the transmutation notebooks and the *Origin of Species* are in fact directed against the teleologists'[1] accounts of organisms and their relations to the external world. But, like most of those who abandoned the teleological approach, Darwin for some time continued to believe in the harmony of nature and in the perfection of adaptation. His adherence to this traditional natural theological vision of nature explains the character of much of his speculation on transmutation from 1837 until the mid-1850s.

The decline of teleological explanation

A cornerstone of the theory of natural selection as it is presented in the *Origin of Species* is the notion of relative adaptation. Forms that are successful in the struggle for existence are deemed to be slightly better adapted than those with which they have had to compete for their places in the economy of nature. But since there is always room for improvement, it cannot be said that they are perfectly adapted for those places.[2] In the mid-nineteenth century, however, most naturalists, including Darwin until the 1850s, believed that adaptation was absolute, not relative; or, in the most commonly used expression, that it was indeed "perfect." This was an opinion sanctioned by the authority of Cuvier and by natural theology. Cuvier supposed that in every type of organism the parts are functionally coordinated, and the whole and all its parts are constructed in the best possible manner for the functions they are to perform and for the situation in which the animal is to live. Life presupposes such fine coordination, as the running of a clock presupposes the clock, he said.[3] Natural theologians, for their part, filled the pages of their works with instances of the adaptation of structure to function and of the whole organism to its environment, for these were supposed to be evidences of purposeful design and hence of an intelligent creator.

For many naturalists adaptation was not merely a fact, but an explanation of facts. It was widely held to be the only acceptable explanation of such general biological phenomena as organic structure, the geographical distribution of plants and animals, and the succession of life on the earth. The eye, the argument went, is constructed as it is because it was made to serve the function of seeing, and the structure of the eye is the best possible structure for that purpose. Similarly, the structure of the whole organism and the time and place of its existence are as they are because of the external conditions to which the organism must be adapted in order to survive. And if fish, reptiles, and mammals have appeared on the earth in succession, it is because these classes were perfectly suited for environmental conditions that appeared successively.

Used in this way, the idea of perfect adaptation constitutes teleological explanation – explanation, that is, in terms of purposes, final causes, or "conditions of existence." The proposition that organisms are to be explained by reference to their conditions of existence received its authoritative formulation in the nineteenth century in Cuvier's *Règne Animal*:

> Natural history, however, also has a rational principle which is peculiar to it and which it uses to advantage on many occasions; it is that of the *conditions of existence*, commonly called *final causes*. As nothing can exist unless it combines in itself the conditions which make its existence possible, the different parts of each being must be coordinated in such a way as to make possible the total being, not only in itself, but in its relations with those which surround it; and the analysis of these conditions often leads to general laws as demonstrable as those which derive from calculation or experiment.[4]

Toby Appel has shown that in Cuvier's subsequent pronouncements, made during the course of a growing controversy with Etienne Geoffroy Saint-Hilaire, this rational principle was transformed into a rigid biological doctrine. For closely related religious and political reasons, Cuvier urged more and more firmly that biological explanation must be teleological.[5] Geoffroy believed that all animals were composed of essentially the same structural elements and that particular organic structures in one animal were to be explained not primarily by the functions they served relative to the life of that animal, but rather by reference to corresponding structures in other animals. The foot of the bear, for instance, is explained, its "general significance" elucidated, when its correspondence with the human hand is demonstrated. The goal of comparative anatomy is to discover the "philosophical resemblances" among organisms.[6] To this Cuvier responded that structural similarities occur only when there is a similarity of function and that the principles employed by Geoffroy to demonstrate the unity of type of animals were simply consequences of the fundamental principle of the conditions of existence.[7]

Cuvier's views were widely adopted in Great Britain, where they were found to harmonize well with the principles of early-nineteenth-century British natural theology. Explanation by function or purpose was central to the most popular version of the argument from design, for it was an organ's fitness for its office that proved it to be the product of intelligence rather than of chance or purposeless mechanical forces. British naturalists lauded Cuvier for his demonstration of the true principles of biological science, and in their own work they generally assumed that

8

organic structure, geographical distribution, and geological succession were to be explained by the teleological principle of
adaptation to the internal and external conditions of life.[8]

It is easy to imagine that Cuvier's prestige, his dominance of
French zoology, and his victory over Geoffroy in 1830 in their
debate before the Academy of Sciences might have combined with
the strong commitment among British scientists to the natural
theological viewpoint as a bulwark of order in a revolutionary
period to make teleological explanation a virtually unchallengea-
ble biological orthodoxy. But in fact, from the 1830s on a growing
number of biologists repudiated teleological explanation in favor
of alternative approaches to the problems of the structure,
distribution, and succession of organisms.[9] Their motives for
doing so were varied and complex.[10] Their arguments, however,
were remarkably similar. Adaptation alone, they agreed, gave an
inadequate account of organisms.

Those who made the shift away from teleological explanation
did not at the same time reject perfect adaptation. Perfect
adaptation had its roots in the belief that organisms are created,
either directly or indirectly, by God, and this belief they retained.
But while they and the teleologists alike believed in perfect
adaptation, they conceived of it differently, and this difference was
reflected in their explanations of biological phenomena. The
teleologists held that the perfection of organisms was due to their
having been specially adapted for particular conditions of existence. They believed, therefore, that perfect adaptation implied a
very close, in a sense causal, connection between the organic and
inorganic worlds, and their theories of organic structure, distribution, and succession were simply theories of adaptation. For
instance, they explained the successive appearance of different
forms of life on the earth as the special adaptation of new
organisms to altered environmental conditions.[11] On the other
hand, those who rejected teleological explanation supposed that
the perfection of organisms was due to their part in a plan of
creation, to their being the result of harmonious laws established
by the creator to achieve his purposes; or in some cases, especially
among the German *Naturphilosophen*, to the purposeful workings
of a world spirit. They denied that perfect adaptation implied any
strict relationship between organic form and inorganic conditions:
the same type of organism might be perfectly suited for many

9

different environmental conditions and many different types suited for the same conditions. Whatever views they held on the source of organic perfection, those who abandoned the notion of strict adaptation were thus in a position to develop theories of organic structure, distribution, and succession which stressed biological relationships and which, as a result, proved generally amenable to recasting in evolutionary form.[12]

One of the clearest statements of the argument against teleological explanation is Richard Owen's *On the Nature of Limbs*. Owen converted the machine analogy so often employed by the teleologists into a clever rebuttal of their principle of adaptation. The teleologists likened the organism to a machine, which, with its parts adapted to particular purposes, shows evidence of contrivance, and therefore of a contriver. The structure of each part is explained by the function which the contriver has designed it to perform. Owen pointed instead to a fundamental dissimilarity between the engineer's machines and the creator's organisms. Man, he said, has made numerous devices for moving on and through the earth, sea, and sky. In doing so he has not made them according to any general pattern or plan, but has adapted them directly to their function. As a result, there are only the remotest analogies among balloons, boats, Stephenson's locomotive, and Brunel's tunneling machinery. The teleologist, Owen said, who wishes to explain organization solely in terms of adaptation to function, should expect to find no greater similarity among the instruments by which various animals traverse these different elements – among, for instance, the forelimbs of man, horse, bat, mole, and dugong. But as every anatomist knows, these limbs, which serve such varied functions, are built according to a single pattern. The goal of the comparative anatomist, Owen said, is to discover the law governing this conformity to pattern. In the search for such law, final causes give us no clue. They are, as Bacon declared, like Vestal Virgins, "barren and unproductive of the fruits we are labouring to attain."[13]

Although *On the Nature of Limbs* appeared only in 1849, the rejection of Cuverian teleology in Great Britain began over a decade earlier, when Owen and a number of other British biologists concluded that Cuvier's principles were inadequate for their needs. An early instance of the shift away from teleological explanation appears in one of the Bridgewater treatises. The authors of the Bridgewater treatises were commissioned to illustrate design in nature, and in most cases they defended teleological explanation as well. Charles Bell in his treatise on *The Hand* (1834), for instance, rejected Geoffroy in favor of Cuvier and insisted that the only principle that should be employed in explaining animal organization is the "principle of adaptation."[14] However, in the same year that Bell's work on *The Hand* was published, Peter Mark Roget brought out the treatise on *Animal and Vegetable Physiology*, in which he suggested that some organic structures cannot be explained solely by their functions, but instead must be referred to a general pattern. Nature, he said, has laid down certain great plans of functions and, in accordance with them, has established general structural patterns for each class of animals. More specialized structures, which carry out the subordinate functions in each species, are not adapted directly to their special functions, but are governed and limited by the general pattern, or type, on which the species is modeled. The student of animal organization must recognize not only adaptation to function – as the teleologist assumed – but also a "law . . . of *conformity to a definite type*."[15]

Two years later, at about the time of Darwin's return to England, Martin Barry argued, somewhat more forcefully, that form, rather than function, was the more fruitful guide to the solution of the most profound problems in natural history. Like Roget, Barry stopped short of declaring the absolute priority of the study of structure:

> It has been usual [he said] to regard organic structure as manifesting design, because it shews adaptation to the function to be performed. It has also been suggested, that function may be equally well considered as the result of structure. And, truly so it may. Yet perhaps we are not required to shew the claim of either to priority; but may consider both structure and function, – harmonizing, as they always do, – as having been simultaneously contemplated in the same design.[16]

Although Barry seems here to be equivocating, he was in fact making the strong claim that as important a consideration as adaptation is, it is not by itself a sufficient explanation of organisms. Throughout the rest of his article he insisted that the investigation of animal organization must focus on structure and development. Adaptations to particular functions merely confuse the naturalist, he said. The deepest insight into nature is to be gained by studying the underlying unity of structure while ignoring the diversity of adaptive features.[17]

In the spring of 1837, Barry's arguments were incorporated by Owen into his first course of Hunterian Lectures at the Royal College of Surgeons, where they formed part of a critique of the teleological approach to comparative anatomy.[18] Barry's article was also read by William B. Carpenter, who was responsible for the most important published critiques of the 1830s. In a review essay in 1838 and again in his widely read textbook of physiology, first published in 1839, Carpenter argued that teleology was not the proper method for the biologist. He was responding to William Whewell's treatment of physiology in his *History of the Inductive Sciences* (1837). In his Bridgewater treatise on *Astronomy and General Physics* (1833), Whewell had insisted that teleological explanation is inappropriate in the physical sciences. Final causes are to be excluded from the investigations of the physical philosopher, he said. Only after the inquiry into the laws governing a phenomenon is complete may the purpose of the phenomenon be considered, at which time the fitness of the means for the end appears as irrefutable proof of the existence of an intelligent creator.[19] But when in his *History* he came to discuss physiology, Whewell characterized Geoffroy's rejection of the teleological method as the "superstition of a false philosophy" and urged that it is both proper and necessary to assume "the existence of an end as our guide in the study of animal organization."[20] Kant had argued that the existence of means apparently designed for particular ends could not be used to prove the existence of God, for it is not possible to know that such arrangements could not be produced by the operation of mechanical laws.[21] Whewell, on the other hand, believed that adaptation to purpose was indeed evidence of creative intelligence. On the method of the biological sciences, however, there was no disagreement. Whewell quoted Kant with approval:

> It is well known that the anatomisers of plants and animals, in order to investigate their structure, and to obtain an insight into the grounds why and to what end such parts, why such a situation and connexion of the parts, and exactly such an internal form, come before them, assume, as indispensably necessary, this maxim, that in such a creature nothing is *in vain*, and proceed upon it in the same

12

way in which in general natural philosophy we proceed upon the principle that *nothing happens by chance*. In fact, they can as little free themselves from this teleological principle as from the general physical one; for as, on omitting the latter, no experience would be possible, so on omitting the former principle, no clue could exist for the observation of a kind of natural objects which can be considered teleologically under the conception of natural ends.[22]

Kant argued not only that the teleological method is necessary in biology, but that its necessity sharply distinguishes biology from the physical sciences. J. D. McFarland has explained that for Kant "any objective explanation must be a mechanical one, and since organisms cannot be explained mechanically, they can never be given an objective explanation."[23] Whewell, who was well aware of the great strides that had been made in the understanding of organic nature since Kant wrote, was not willing to go so far.[24] But his treatment of physiology makes clear that he assigned it a status inferior to that of physics.

To this, Carpenter, reviewing Whewell's *History*, objected strongly. Physiology, he declared, has the same claim to be considered an "inductive science" as does physics or astronomy. The problem with physiology is not that it is essentially different from physics, but that in physiology it is so much more difficult than in physical science to discover mere facts, much less to derive laws from them. Whewell's examples of great *"doctrines"* which the teleological method had established – such as the circulation of the blood – are not doctrines, or *"laws,"* at all, Carpenter said. They are rather some of the facts whose discovery is necessary before general laws can be formulated. "Were we able to ascertain *facts* regarding the changes which take place in the interior of the living body as easily as the astronomer observes the place of a planet, or the chemist the decomposition of a salt, there is no reason whatever to prevent these facts being generalized in the same manner and to the same degree with those of the physical sciences."[25]

Carpenter went on to criticize Whewell's discussion of the dispute between Geoffroy and Cuvier. Geoffroy's rejection of final causes, Carpenter said, and his statement "I ascribe no intention to God" are no more irreligious than Whewell's prescription of method for the physical sciences. "We are not to assume that we know the objects of the Creator's design, and put this assumed purpose in the place of a physical cause," Whewell had written.[26] Were biologists to follow Whewell's advice and adhere to the method of final causes, as practiced by Cuvier, their science, Carpenter said, would be limited to the discovery of facts, while general laws would be lost from view:

> The philosophic Physiologist, who is not deterred by the clamour of bigotry and prejudice, will follow precisely the same course [as the physical philosopher]. The adaptation which he discovers in particular instances may well serve both to awaken his curiosity, and to lead him to suspect a pre-existing Design. But he will obtain a much more elevated view of the nature of Creative power, if he carry his enquiries further. He must disregard for a time, as in physical philosophy, the immediate *purposes* of the adaptations which he witnesses; and must consider these adaptations as themselves but the *results* or *ends* of the general laws for which he should search.[27]

Carpenter's statement of the method of physiological research expresses well the views of those biologists who were dissatisfied with the method of final causes. With few exceptions, if any, they believed that there was purpose in nature and that structure was admirably adapted to function. They did not hesitate to cite adaptation as a proof of design or to explain particular structural modifications in functional terms. But they denied that function alone accounts for structure, and they held that the physiologist was not to rest content with the discovery of purposes. His goal should be rather to find the "general laws" that governed both structure and its adaptation to function.[28]

In the first edition of his *Principles of General and Comparative Physiology*, Carpenter extended his criticism to the teleologists' approach to the question of organic succession. During the 1830s and 1840s most British geologists, and the older biologists, such as Charles Bell, offered a teleological explanation of the appearance of new species. Central to this explanation was the "principle of adaptation," which stated that every organism is perfectly adapted to the situation that it occupies in the economy of nature. As new environmental conditions have arisen during the history of the earth, new species, specially suited for these conditions, have been created by some unknown means. According to this view, the succession of organisms on earth is wholly dependent on the development of the earth itself. This teleological interpretation of the history of life is most clearly displayed in Charles Lyell's writings, for Lyell made it into a central element in the antievolutionary strategy of his *Principles of Geology* (1830–3).[29] Later, when he began to reconsider his stance on evolution, Lyell set down in his private species journals an admirably explicit statement of the assumptions on which he and his fellow geologists had worked. For Lyell, perfect adaptation implied environmental determinism, a determinism in which the "necessity" of the relationship between organism and environment lay in the goodness of God. Of all the ways, Lyell said, in which the creator might adapt a new organism to the conditions under which it and its descendants are destined to live, it may be that one way "is preferable to all others." In this case, the creator will always choose that one. For every given set of environmental conditions, there will be called into existence that organic form which is the best suited of all possible forms for those conditions. Since only one form can be the best of all possible forms, external conditions determine which forms are created. "What is here called necessity," Lyell said, "may merely mean that it pleases the Author of Nature not simply to ordain fitness, but the greatest fitness."[30]

Lyell no doubt recognized that it was not possible to demonstrate in any satisfactory fashion that a given species is the best possible for the conditions in which it exists. But he thought it could be proved that a deterministic and uniform law of adaptation is at work by showing that similar conditions always "produce" similar forms of life.[31] For support Lyell turned to the study of geographical distribution, and in particular to the writings of De Candolle. It was well known, and De Candolle emphasized, that widely separated regions of the earth which have similar climates are not generally inhabited by the same species of plants and animals. It would probably not be difficult, he said, to find two points in the United States and Europe, or America and equinoctial Africa, in which the same external conditions exist – the same temperature, elevation, soil, and humidity – but in which all, or nearly all, the plants are different. Such facts were not at all favorable to Lyell's environmental determinism. But De Candolle went on to say that there is nevertheless a strong analogy of form between inhabitants of regions in which external conditions are similar. If the species are not the same, the genera often are.[32] It was in this statement that Lyell found justification for his proposition that similar conditions always produce similar forms of life.

15

De Candolle himself concluded that the differences separating the species which inhabit areas with similar conditions indicate that some factors besides adaptation to conditions are involved in determining the distribution of organisms.[33] Although Lyell quoted De Candolle's opinion on this point, he did not accept it. Lyell believed that the principle of adaptation alone could account for all the phenomena. De Candolle wrote that "a portion of the phenomena of the distribution of plants in the different countries appears to relate to the appreciable influence of temperature; but there remain some facts which elude all existing theories because they relate to the very origin of organized beings, that is to say, to the most obscure subject in natural history."[34] Lyell, on the contrary, said that if similar conditions do not produce the same species, this is due not to any additional, as yet unknown, biological laws, as De Candolle seemed to suggest, but to the fact that the climates, though similar, are not identical. The environmental conditions that in different regions or at different geological periods determine the existence of particular species are too complex ever to be precisely duplicated, he said. But they may be so nearly duplicated as to produce the same *genera* in different regions or at different times in the past and future. For Lyell the teleological explanation of organic succession was sufficient.[35]

In other respects, Lyell's opinions are not characteristic of the geological thought of his day.[36] Lyell argued that the history of the earth is cyclical and that the history of the organic world is a parallel series of cycles: When the current cycle of an earlier geological age returns, then iguanodons, ichthyosaurs, and pterodactyls, which were perfectly suited to the ancient climate, will again inhabit the earth.[37] Adam Sedgwick, William Buckland, Henry De la Beche and other progressionists supposed that the earth has undergone a directional, rather than a cyclical, development and that at successive geological epochs it has been suited to the existence of "higher" forms of life.[38] On the questions of directional development and organic progress, there was indeed a fundamental disagreement between the uniformitarian Lyell and most of his colleagues. But on the question of the explanation of organic succession there was an equally fundamental agreement. The progressionism of the majority of British geologists was built on the same foundation as Lyell's nonprogressive cycles of life – the principle of strict adaptation to changing conditions. The progressionists were as little willing as Lyell to admit the insufficiency of the teleological explanation of the history of life. Progress, they claimed, depends not on some biological law of development, but on progressive changes in the environment. "There have been successive creations as new conditions have arisen, so that every place capable of sustaining life has been filled by that fitted for it," wrote De la Beche. Animals and plants, Buckland said, "were constructed with a view to the varying conditions of the surface of the Earth, and to its gradually increasing capabilities of sustaining more complex forms of organic life, advancing through successive stages of perfection." And Sedgwick insisted that organic succession was to be explained not by some biological law but by "creative power" adapting new forms of life to gradually changing conditions.[39] Adaptation, for them as for Lyell, was the sole explanation for change.

To Carpenter, Owen, and Louis Agassiz, among others, this teleological explanation of organic succession was no more acceptable than the teleological explanation of organic structure. Carpenter in 1839 suggested that "those who have dwelt most upon this adaptation of the structure of living beings to the external conditions in which they exist, appear to have forgotten that these very conditions might be regarded, with just as much propriety, as specially adapted to the support of living beings."[40] In one sense Carpenter's charge is untrue. Lyell and the progressionists both viewed the organic and inorganic worlds as mutually adapted, as two halves of one harmonious plan. But for all practical purposes Carpenter was right. The teleologists failed to see or were unwilling to admit the implications of the possibility that the earth might be "specially adapted" for the existence of living beings. They did indeed believe that external conditions were designed with the organic world in mind. But they supposed that the external conditions at any period, though known beforehand by God, were the product of geological forces and laws, not of special adaptation to the living beings then or about to be in existence. The fitness of the conditions for the organisms resulted from the fact that the development of the inorganic world was planned by the creator so as to be suitable for the successive forms of life. Organisms, on the other hand, were supposed to be "specially adapted" to conditions that gradually arose during the earth's history. The geologists always treated geological change as primary and organic change as dependent upon it. In their eyes there was a strict parallel between the history of the earth and the history of life, and, most importantly, designed adaptation to conditions was a wholly sufficient explanation of the organic half of the parallel.[41]

For Carpenter, adaptation was no explanation at all, but was rather the result of laws that remained to be discovered. What he was saying is that it is possible that events in the *organic* world, though known beforehand by God, are the result of biological forces or laws, not of special adaptation to conditions. It is as reasonable to suppose that biological change is primary, and that geological changes are "specially adapted" to it, as the reverse. And since one supposition is as reasonable as the other, the notion of special adaptation has no explanatory value in either geology or biology. The student of living beings, like the student of external conditions, must ignore adaptation and search instead for the underlying laws and forces that produce it.

In their writings in the 1840s and 1850s Owen and Agassiz adopted positions similar to Carpenter's. Like Carpenter, Agassiz in his *Essay on Classification* criticized both the method of final causes in comparative anatomy and the teleological explanation of organic succession. The explanation of organic structure in terms of the functions that the organs are intended to serve is the basis for most general works on comparative anatomy, he said, "and yet there never was a more incorrect principle, leading to more injurious consequences, more generally adopted." As proof of the inadequacy of the functional approach, Agassiz cited the fact that throughout the animal kingdom identical functions are performed by a great variety of organs.[42] Agassiz argued that similar facts made untenable the teleologists' interpretation of perfect adaptation to external conditions. Adaptation is indeed perfect, he said, and its perfection is evidence of creative wisdom and power. But there does not exist any strict relationship between particular organisms and particular external conditions of existence. On the contrary, organisms exhibit an extraordinary independence from the influence of physical agents. Much of the first chapter (some two hundred pages) of Agassiz's *Essay* is devoted to refuting environmental determinism, and this is the sole aim of sections 2 and 3 – "Simultaneous existence of the most diversified types under identical circumstances" and "Repetition of identical types under the most diversified circumstances." The creator, Agassiz insisted, has adapted many different types of organisms to the same conditions and has adapted the same type to many different conditions. He denied absolutely the teleologists' interpretation of perfect adaptation. Environmental determinism and the explanation of organic succession simply in terms of adaptation to the developing inorganic world were, in his eyes, absurd.[43]

Agassiz, Carpenter, and Owen shared with teleologists like Lyell the view that organisms are perfectly adapted to conditions, and they agreed that geological change could cause the extinction of some forms and call for the introduction of others that were adapted to the new conditions. Owen, who accepted Lyell's gradualist view of the history of the earth, treated organic succession as a response to environmental change, as did Lyell. But for Owen there was no strict relationship between external conditions and organic form; there was no single best form for a given set of conditions. He believed that the structure of successively appearing organisms was determined not only by the conditions to which they must be adapted, but also by biological laws, such as unity of type and the law of the succession of the same type in the same country. He supposed that new forms were related not only to conditions, but also to their predecessors.[44]

Their recognition of this kind of biological relationship made the antiteleologists' views compatible with transmutationism, and this placed them in a delicate situation. One reason why the principle of adaptation had seemed so attractive to many naturalists in the first half of the nineteenth century was its antievolutionary implications. Teleological explanation did not make all naturalistic theories of the introduction of new organic forms impossible, but it did rule out theories of descent. In insisting that organic structure is completely accounted for by adaptation, the teleologist excluded all other possible determinants of structure, such as heredity. Those who rejected the principle of adaptation thus laid themselves open to the charge that they were, perhaps unwittingly, lending support to proponents of the transmutation of species.[45] The issue of biological explanation was a sensitive one, and consequently naturalists like Owen, Carpenter, and Agassiz were anxious to explain that they did not see their viewpoint as tending to denigrate God's power or to minimize God's influence in the creation. On the contrary, they asserted, they were contributing to the progress of natural theology as well as of biological science. Agassiz saw himself as laying a new and better foundation for natural theology. He wished to rid it of the unsophisticated "proofs" of the existence of God which characterized the Bridgewater treatises and to found it instead on the most recent improvements in biology. The adaptation of means to ends, he said, is not only an inadequate explanation of organic structure, but also an inadequate proof of the existence of an intelligent creator. "For we can conceive," he said, "that the natural action of objects upon each other should result in a final fitness of the universe, and thus produce an harmonious whole." What it is not possible to conceive, Agassiz argued, is that mere matter in motion could produce the vast intellectual scheme whose existence biological investigation is daily uncovering. Biologists contribute most effectively to natural theology not by discovering more examples of the adaptation of structure to function and organism to environment, but rather by discovering the laws that constitute the creator's plan. In defense of his antievolutionary position Agassiz insisted that these laws themselves prove that organized beings owe their successive appearance on the earth to the "immediate action of a thinking being."[46]

On the question of whether creation was due to divine intervention or to the continuous operation of natural causes, Owen and Carpenter disagreed with Agassiz. By the 1840s Owen thought that, rather than divine intervention, the facts of morphology and paleontology suggested that the introduction of new species was due to "natural laws or secondary causes."[47] And Carpenter supposed that the laws that define the plan of creation were subordinate to one "simple law," impressed on matter in the beginning, which brought about the creation of the universe and the creation and succession of life, all in accordance with a divine plan.[48] In their discussions of natural theology, however, Owen and Carpenter were quite close to Agassiz. Carpenter, like Agassiz, urged that the natural theologians relied on a weak argument for the existence of God. Adaptation, he said, might be the result of something like natural selection rather than design. The best proof of creative intelligence, he thought, lay in the "comprehensive plan" of creation, which could only be the product of "Infinite Wisdom." Only by rejecting the teleological method, he said, is the naturalist able to discover the creator's plan.[49]

In his Hunterian lectures for 1838 Owen likewise argued that in deviating from the teleological method in comparative anatomy there was no danger of losing sight of the fact that organisms owe their existence to creative wisdom. Law, he said, is as clear a sign of intelligence as adaptation:

> It is a groundless fear that some have expressed when a physical explanation of the form and condition of an organ has been offered, that it blunts our appreciation of the Wisdom which adapted that organ to its special office. On the contrary – the higher and more general are the laws regulating the structure of animals of which we can obtain a conception, the more will the Contempletative Mind be struck with the vastness of that designing intelligence which in originally ordaining them could produce such harmony and adaptation amongst their innumerable results.[50]

In *On the Nature of Limbs* he took the same line and extended it further. It is true, he conceded, that the rejection of final causes in favor of such laws as conformity to a general pattern or type implies that in one sense some organic structures are "made in Vain" – that is, not made for a particular, identifiable purpose – a repugnant proposition from the point of view of natural theology and one which, according to Whewell and Kant, contradicts the fundamental postulate of biological science. But it is the study of precisely those organs and structural patterns that cannot be explained by their functions which leads to the discovery of important truths about the organic world. Such structures are not made in vain, Owen said, if their "truer comprehension lead rational and responsible beings to a better conception of their own origin and Creator." He went on to urge that his formulation of the law of unity of type refuted atheism. The Greek atomists, he said, argued that if the world was made by a deity, then the idea of the world must have existed before the world itself, a notion which they dismissed. Owen believed that his discovery of the vertebrate archetype proved that ideas indeed existed before things: "The recognition of an ideal Exemplar for the Vertebrated animals proves that the knowledge of such a being as Man must have existed before Man appeared. For the Divine mind which planned the Archetype also foreknew all its modifications."[51]

Owen, Carpenter, and Agassiz were all saying very much the same thing, and it is not difficult to find other examples of biologists who adopted a similar point of view at about the same time. Theodor Schwann, for example, argued in almost the same terms as Carpenter employed that there is as little room for teleological explanation in biology as there is in the physical sciences. Teleological explanations in biology, Schwann urged, are of no more value than it would be to say that

> the motion of the earth around the sun is an effort of the fundamental power of the planetary system to produce a change of seasons on the planets, or to say, that ebb and flood [of the tides] are the reaction of the organism of the earth upon the moon.
>
> In physics, all those explanations which were suggested by a teleological view of nature, as "horror vacui," and the like, have long been discarded. But in animated nature, adaptation – individual adaptation – to a purpose is so prominently marked, that it is difficult to reject all teleological explanations. Meanwhile it must be remembered that those explanations, which explain at once all and nothing, can be but the last resources, when no other view can possibly be adopted; and there is no such necessity for admitting the teleological view in the case of organized bodies.[52]

Like Carpenter in his review of the *History of the Inductive Sciences*, Schwann rejected the idea that biology must employ a method that is essentially different from that of the physical sciences. But he believed, nonetheless, that organized beings, and the inorganic world as well, show adaptation to purpose. "We know, for instance, the powers which operate in our planetary system," he said. "They operate, like all physical powers, in accordance with blind laws of necessity, and yet is the planetary system remarkable for its adaptation to a purpose. The ground of this adaptation does not lie in the powers, but in Him, who has so constituted matter with its powers, that in blindly obeying its laws it produces a whole suited to fulfil an intended purpose."[53] The organic world too, with all its adaptations, may well be the result of laws established by God in the beginning.

To Schwann, Agassiz, Owen, and Carpenter – as well as to Henri Milne Edwards, Karl Ernst von Baer, Edward Forbes, and indeed, most of the leading naturalists of the middle third of the century – it appeared that the facts of their science were incompatible with a strictly teleological interpretation of organisms.[54] But however inadequate or philosophically unsatisfying teleological explanations seemed to them to be, they generally remained committed to the belief that the universe exhibits both order and purpose. They spoke of a "teleology of a higher order," by which they usually meant a purposeful system of laws that produce both the perfection of individual organisms and the evidence of plan and foreknowledge.[55] Their rejection of teleological explanation did not entail at the same time a rejection of traditional ideas about the harmony and overall perfection of nature. These they continued to share with the teleologists and natural theologians whose approach to biological problems they repudiated. It might be said that these ideas formed a common set of assumptions that served to link the numerous competing philosophies of nature in the period – the common English Christian view that the laws of nature are an expression of God's immediate and perpetual action; the in many respects similar deistic view that God created matter and the laws of its motion and left them to achieve his ends; and pantheistic notions of the purposeful activities of the world spirit. An antiteleologist, as well as a teleologist, might in principle have subscribed to any of these philosophies, or to some other. In other words, there occurred a shift in styles of biological explanation without this having any decisive impact on the competition among different philosophies of nature and without its implying any departure, at least in the opinion of the antiteleologists, from those assumptions of perfection, harmony, and cosmic purpose that were shared by most naturalists. As I have just indicated, however, it soon became apparent – and to Darwin as soon as to Carpenter or Owen – that the shift had significant implications not only for theoretical natural history, but also for the arguments, especially the antievolutionary arguments, of natural theology.

Darwin, teleological explanation, and the harmony of nature

There is a close parallel between the shift in biological opinion which I have just described and the changes that occurred in Darwin's views from the period of the *Beagle* voyage until the fall of 1838, when he first formulated the theory of natural selection.[56] Darwin began by believing in perfect adaptation in the teleologists' sense, employing the notion of strict adaptation to explain both organic distribution and succession. Then, concomitant with his acceptance of transmutation, he, like so many of his contemporaries, rejected teleological explanation, saying that adaptation alone could not explain the phenomena of organic nature. But through the first fifteen months of his notebook speculations on transmutation, Darwin retained and explicitly stated his belief that adaptation is perfect, one result of the harmonious system of laws established by the creator.

The most easily accessible source of information on Darwin's early views is his autobiography. There he described the pleasure he derived from the study of Paley's works, including *Natural Theology*, and said that he had once found "the old argument of design in nature, as given by Paley, . . . conclusive."[57] These brief recollections are little more than suggestive, however, for they give no indication of how the viewpoint of Paley's natural theology governed Darwin's interpretation of biological phenomena. For this we have to turn to notes he made during the voyage, in which can be found several passages that reveal Darwin's views on the harmony of nature and a few that show him giving teleological explanations of the succession and distribution of organisms.[58] Some of the former are fairly well known, such as the discussion of ant lions in the diary of the voyage, with its reference to the beauty of the ant lion's trap and to the completeness of the creator's work.[59] Much more explicit statements of the traditional conception of the harmony of nature occur in Darwin's "General Observations" on the zoology of Rio de Janeiro (visited April–June 1832) and Maldonado (visited May–June 1833):

> I could not help noticing how exactly the animals & plants in each region are adapted to each other. – Everyone must have noticed how Lettuces & Cabbage suffer from the attacks of Caterpillars & Snails. – But when transplanted here in a foreign clime the leaves remain as entire as if they contained poison. – Nature when she formed these animals & these plants knew they must reside together.[60]

The absence of dung beetles at Maldonado, despite the presence of herds of horses and cattle, occasioned this reflection:

> This absence of Coprophagous beetles appears to me, to be a very beautiful fact; as showing a connection in the creation between animals so widely apart as Mammalia & Insecta Coleoptera, which when one of them is removed out of its original zone can scarcely be produced by a length of time & the most favourable circumstance . . . If proofs were wanting, to show the Horse or Ox to be aboriginals of great Britain I think the very presence of so *many* species of insects feeding on their dung, would be a very strong one.[61]

In notes that appear to date from 1835 and 1836, by which time Darwin was a self-conscious disciple of Lyell, environmental determinism is invoked to explain the distribution of organisms and the appearance of new forms. As Lyell had done, Darwin supposed that the fixed relationship between particular environmental conditions and the existence of certain genera made piecemeal extinction and appearance of forms intelligible: as conditions gradually change, plants and animals adapted to the former state of affairs must perish while others are created to take their places:

> As Mr. Lyell supposes species may perish as well as individuals; to the arguments he adduces, I hope the Cavia of B. Blanca will be one more small instance, of at least a relation of certain genera with certain districts of the earth. This co-relation to my mind renders the gradual birth & death of species more probable.[62]

By "gradual" Darwin must be understood to mean that species come into existence one or a few at a time, rather than in complete new creations, for he conceived the appearance of single species to be sudden, even instantaneous:

> We shall presently run over the proofs of repeated elevations: May we conjecture that these began with greater strides, that rocks from sea too deep for life were rapidly elevated & that immediately when within a proper depth life commenced . . . The elevations rapidly continued; land was produced, on which great quadrupeds lived: the former inhabitants of the sea perished (perhaps as effect of these changes) the present ones appeared.[63]

This is the clearest instance I have seen, not excluding the works of Lyell, Buckland, or Bell, of the use of the teleological approach in solving a specific scientific problem: new forms appear as soon as the appropriate conditions occur. Darwin did not consistently assume that the introduction of new forms follows "immediately" on the establishment of the set of conditions that makes them possible, as is evident from his explanation, two pages further on in the same manuscript, of the absence of trees in the fertile Pampas. This he attributed to "no Creation having taken place subsequently to the formation of the superior Tosca bed."[64]

In the winter or early spring of 1837, a few months after his return to England, Darwin became a transmutationist, a decision based apparently on careful consideration of the implications of observations made during the voyage and of the results of examinations by experts of some of his zoological specimens. Notebook passages written at this time indicate he was immediately aware that as a system of explanation the teleological approach of Lyell and Cuvier was incompatible with his new beliefs.[65] Darwin's private debate with the teleologists runs through the whole series of transmutation notebooks he began in July 1837. Often it takes the form of an argument against "creation." By "creation" Darwin meant not so much the notion of direct intervention by the creator as the idea of special adaptation. It made comparatively little difference whether "creation" was by secondary means, as Lyell seemed to assume, or by immediate intervention, as Buckland and Sedgwick probably believed; in Darwin's most common usage, what constituted the hypothesis of "creation" was the belief that each form is fully explicable in terms of its own needs and purposes.[66] Here is a straightforward example of such usage: Darwin argued that the wings of flightless beetles cannot be explained by the teleological assumption that every organ is of some functional importance to its possessor. These wings show, he reasoned, that the beetles were "born from beetles with wings and modified, – if simple creation, surely would have [been] born without them." It is not the mode of introduction, but the teleological assumption, that is at issue. If organisms are formed solely with reference to their conditions of existence, then heredity, and hence transmutation, are left out of the question.[67]

Darwin's attack on teleology began early in the first notebook, when he offered his theory of "propagation of species" as a solution to problems left unresolved by the teleological account of geographical distribution. Lyell had argued that external conditions, together with the means of migration and dissemination of seeds bestowed on each species at the time of its creation, could fully account for the phenomena of geographical and geological distribution. But Lyell's principal source of information on this question, De Candolle, urging that environmental determinism was an inadequate explanation, had said that there also come into play certain laws which appear to relate to the origins of things and of which we are ignorant. Propagation, Darwin said, fills this gap in our knowledge: "On this idea of propagation of species we can see why a form peculiar to continents, – all bred in from one parent . . . This is answer to Decandolle . . . : genera being usually peculiar to same country, different genera – different countries." Where the teleological argument fails, or is unconvincing, propagation offers a plausible alternative. It shows why certain organic structures are peculiar to certain countries when we can hardly believe they are necessary for purposes of adaptation: "If it was necessary to one forefather, the result would be as it is."[68]

Sometimes Darwin turned the teleologists' own arguments against them, as Owen did with the machine analogy. For instance, Lyell, in common with the majority of naturalists after Linnaeus, supposed that each form of life has originated in only one place. If it is now found in two widely separated places, he assumed that it was created in one and transported to the other, and he concluded that the many means that exist for the dissemination of animals, plants, and their seeds are given them for the purpose of introducing each form of life into every region where it is destined to live and for which it is specially adapted.[69] Darwin, on the other hand, saw in the means of dissemination discussed by Lyell an indication that animals are not in fact specially adapted for particular countries. Rather, their numerous mechanisms of dispersal allow them to reach as many potential habitats as possible.[70]

Darwin's own observations and his reading provided him with numerous other facts that seemed to him to be inexplicable by the teleologist. If organisms come into existence when the external conditions suited to them arise, why, he asked, are some animals not found in places where there are manifestly appropriate conditions for them? Why are large mammals not found on islands? "If act of fresh creation, why not produced on New Zealand; if generated, an answer can be given." Conversely, why are similar forms found under different conditions; for instance, plants that grow in the volcanic soils of islands but that are of the same type as those growing in sandstone and granite soils in Africa? And there are species of molluscs that are common to Patagonia, to Tierra del Fuego, and to forest, which shows "independency of shells to external features of land." Still other facts lead to the same conclusion. Races of domestic animals are shaped by the environmental influences of the country in which they originate. Yet they sometimes flourish in other countries where the conditions are different. This indicates that species are not "so closely adapted" as the teleologists suppose.[71] We who always have in mind Darwin's achievements in the *Origin of Species* tend frequently to assume that the lines of thought he developed in his notebooks and later pre-*Origin* writings were as exclusively

Darwinian as the theory of natural selection. It is a useful corrective to recall that such opponents of transmutation as Prichard and Agassiz used these same arguments to reach the same conclusion – that the principle of adaptation could not explain geographical distribution.[72]

The explanation of organic structure offered by Cuvier and the natural theologians was also unacceptable to Darwin, for it suffered from the same defects as teleological explanations of geographical and geological distribution. Adaptation might well explain some structures, but it could not explain them all. It could not explain, for instance, vestigial or rudimentary organs which served no apparent purpose in their possessors; it could not explain the unity of type of such large groups as the vertebrates; and it could not explain the similarities between fossil and living inhabitants of the same country, except in those cases where external conditions had not changed. If, however, the principle of adaptation is supplemented by another principle, heredity, then such phenomena make sense. "The condition of every animal," Darwin said, "is partly due to direct adaptation and partly to hereditary taint; hence the resemblances and differences, for instance, of finches of Europe and America, etc. etc. etc."[73]

It appears that Darwin first became aware of the conflict between the proponents of the teleological and nonteleological approaches to problems of organic structure in 1837, when he read Geoffroy's account of his debate with Cuvier. Darwin was not convinced by Geoffroy's arguments for the unity of type of all animals ("I deduce from extreme difficulty of hypothesis of connecting mollusca and vertebrata, that there must be very great gaps"). And he could not figure out what Geoffroy's views were on several important issues. But Darwin nevertheless saw that Geoffroy's ideas were more congenial to his own than were Cuvier's. He was so sure of this that he assumed, without it seems knowing it to be true, that Geoffroy believed in descent: "does not say [animals] propagated, but must have concluded so." In Geoffroy's copious extracts from Cuvier's writings, on the other hand, Darwin noted only the inadequacies of Cuvier's explanations. Cuvier's notion that each animal is constructed with reference solely to its own needs, and not to any general type form, "does not agree with old and modern types being constant," Darwin said. (Already in the opening pages of the first notebook he had written that "propagation explains why modern animals – same type as extinct, which is law [of 'propagation'] almost proved.")[74] Nor could Cuvier's "conditions of existence" explain the fact that each class of vertebrates, for instance, has representatives adapted for many different situations. But again, heredity plus adaptation might do so: "Perhaps consideration of range of capabilities past and present might tell something."[75]

For Darwin, as for others, rudimentary organs were among the strongest arguments against the teleologist, for only with great difficulty could they be construed as serving some useful purpose. Structures which seem to be unnecessary, and are therefore inexplicable by the teleologist, can be accounted for by inheritance from an ancestor to whom they were of some use: "armadillos like every created [edentate]. – Passage for vertebrae in neck – same cause; such beautiful adaptation, yet other animals live so well. – This kind of propagation gives hiding-place for many unintelligible structures – it might have been of use in progenitor, or it may be of use, – like mammae on men's breast. – " When one sees a nipple on a man's breast, Darwin had said a few pages earlier, one must say not that it has some use, but that it is a vestige of a hermaphrodite ancestral form.[76]

Darwin rightly saw that the teleological approach to structure and distribution was a rival to his own view that species are "propagated," and he seized on every fact or argument that cast doubt on the associated ideas of a fixed relationship between organisms and their environments and of the functional necessity of every detail of organic structure. Taken together those facts and arguments seemed to Darwin to require the rejection of the whole explanatory system of Cuvier and the natural theologians, including their bulwark against transmutationism: "Look abroad, study gradation, study unity of type, study geographical distribution, study relation of fossil with recent. The fabric falls!"[77]

In his dissatisfaction with the narrow interpretation of perfect adaptation, Darwin was in agreement with most of the best young biologists of his day. If this were the only similarity in their views, it could perhaps be dismissed as an unimportant coincidence. Darwin was bound to reject teleological explanation because of his belief in transmutation. Owen, Carpenter, Agassiz, and Schwann, on the other hand, rejected it on the ground that it gave an unsatisfactory account of some of the principal phenomena of their (nontransmutationist) science. (Of these four, only Owen seems to have believed in descent before 1859.) But the correspondence between Darwin's thought and that of his contemporaries is not confined to their rejection of teleological explanation. It extends to many of the views they held, as well as those they repudiated, and from this it appears that during the period of the first three transmutation notebooks, Darwin was operating within the same general framework of beliefs about the harmony and overall perfection of nature as were most of his fellow naturalists. Many abandoned the narrowly teleological viewpoint. But the laws, of whatever sort, that they substituted for particular explanations in terms of purpose were themselves thought to be purposeful. Each law was supposed to play its part in carrying out the creator's grand design. This is the "teleology of a higher order" of which Owen, and later Whewell and Huxley, spoke, and it is a teleology that Darwin continued to accept long after he became a transmutationist.

Darwin's notebooks are permeated with the conviction that the "Creator creates by . . . laws," that each law serves a purpose in the whole system of laws, and that this is a far higher conception of the creator than to suppose he creates by separate interpositions of divine power.[78] In these convictions and in the form Darwin's notebook speculations took, there is a marked resemblance to the sentiments of Owen and Carpenter quoted earlier. Carpenter exhorted biologists to ignore purposes and look instead for general laws. After the laws are discovered, then their purposes may be inquired into. This was very nearly the course Darwin followed in the notebooks. His interest focused not on particular purposes, but on general relationships. And for each law to which his speculations led him, he posed the question of its purpose. It is a question that recurs again and again: What is the final cause of this? Most frequently it was asked of the development of sexual generation, which, as David Kohn has shown, occupied a central position in Darwin's early theories.[79] To this Darwin gave several answers. One, of which more must be said shortly, is that it is necessary to adapt species to changing circumstances: "why two sexes – not necessary to generation . . . I can scarcely doubt final cause is the adaptation of species to circumstances by principles, which I have given."[80] Another is that it is necessary for the production of social animals, especially man:

My theory gives great final cause of sexes in separate animals: for otherwise there would be as many species, as individuals [because it is sexual generation that maintains specific constancy by blending out differences due to minor external changes], & though we may not trace out all the ill effects, – we see it is not the order in this perfect world, either at the present, or many anterior epochs. – but we can see if all species, there would not be social animals, hence not social instincts, which as I hope to show is probably the foundation of all that is most beautiful in the moral sentiments of the animated beings . . .

Whether he was or not He [man] is [at] present a social animal. If man is *one* great object for which the world was brought into present state, – a fact few will dispute, (although, that it was the sole object, I will dispute . . .) & if my theory be true then the formation of sexes rigidly necessary.[81]

The assumption underlying most of Darwin's inquiries into final causes in the notebooks is that expressed here – that this is a "perfect world," its order the result of laws established to achieve certain ends.

This conception of law and purpose had been common in physics and astronomy since the time of Boyle and Newton. Schwann and Carpenter, in passages quoted earlier, argued that it ought to be extended to biology as well. This was Darwin's point also in those places in the notebooks in which he compared the law of gravity and the law of transmutation.[82] In his most general statements of his conception of a system of laws, the agreement between Darwin's viewpoint and that of Carpenter, Schwann, Owen, and others is unmistakable. In the first edition of his *Principles of General and Comparative Physiology*, Carpenter, in concluding both the book and his criticism of Whewell, wrote:

It has already been remarked, that the rapid progress of generalisation in the physical sciences renders it probable that, ere long, a simple formula shall comprehend all the phenomena of the inorganic world; and it is not, perhaps, too much to hope for a corresponding simplification in the laws of the organised creation, although its progress is necessarily retarded by the many obstacles, which the nature of the subject presents to the philosophic enquirer. Every step which we take in the progress of generalisation, increases our admiration of the beauty of the adaptation, and the harmony of the action, of the laws we discover; and it is in this beauty and harmony that the contemplative mind delights to recognise the wisdom and beneficence of the Divine Author of the Universe . . .

If, then, we can conceive that the same Almighty *fiat* which created matter out of nothing, impressed upon it one simple law, which should regulate the association of its masses into systems of almost illimitable extent, controlling their movements, fixing the times of the commencement and cessation of each world, and balancing against each other the perturbing influences to which its own actions give rise, – should be the cause, not only of the general uniformity, but of the particular variety of their conditions, governing the changes in the form and structure of each individual globe, protract-ed through an existence of countless centuries, and adjusting the alternation of "seasons and times, and months and years," – should people all these worlds with living beings of endless diversity of nature, providing for their support, their happiness, their mutual reliance, ordaining their constant decay and succession, not merely as individuals but as races, and adapting them in every minute particu-lar to the conditions of their dwelling, – and should harmonise and blend together all the innumerable multitude of these actions, making their very perturbations sources of new powers; – when our knowledge is sufficiently advanced to comprehend these things, then shall we be led to a far higher and nobler conception of the Divine Mind than we have at present the means of forming.[83]

At very nearly the same date,[84] Darwin made the following entry in his third transmutation notebook:

16th Aug. [1838] What a magnificent view one can take of the world Astronomical causes modified by unknown ones, cause changes in geography & changes of climate suspended to change of climate from physical causes, – then suspended changes of form in the organic world, as adaptation, & these changing affect each other, & their bodies by certain laws of harmony keep perfect in these themselves. – instincts alter, reason is formed & the world peopled with myriads of distinct forms from a period short of eternity to the present time, to the future. – How far grander than idea from cramped imagination that God created (warring against those very laws he established in all organic nature) the Rhinoceros of Java & Sumatra, that since the time of the Silurian he has made a long succession of vile molluscous animals. How beneath the dignity of him, who is supposed to have said let there be light & there was light.[85]

Certainly not all biologists would have been willing to say that such laws could be considered the efficient causes of new species, as Carpenter and Darwin implied. Yet all would have agreed that they were parts of a purposeful system.

The laws having been established by God, their consequences

must be harmony and perfection. Of this, Darwin had no doubt. His conviction on this score is expressed numerous times in the transmutation notebooks: "how nicely things [are] adapted"; the object of laws of organic change is to produce "complete adaptation"; by "laws of harmony" organisms "keep perfect"; organisms are "perfectly adapted" to the circumstances and times of their existence; geological and organic changes are parts of a "system of great harmony."[86] Except for the reference to laws of change, these sentiments are little more than reiterations of the views that we find in the *Beagle* notes, before Darwin was a transmutationist. Prior to the autumn of 1838, when he formulated the theory of natural selection, Darwin's evolutionary speculations had forced him to reject teleological explanation. But they had not led him to question the traditional conception of the harmony of nature, any more than his contemporaries' similar rejection of the method of Bell and Cuvier had led them to abandon the harmonious view. Like them, Darwin continued to believe in a variety of perfect adaptation. And, like Owen's law of unity of type, Darwin's "law of transmutation" was simply one part of the harmonious system of creation.

Perfect adaptation

In the writings of Darwin and the biologists of his generation it is possible to distinguish, by the late 1830s, between two varieties of the concept of perfect adaptation. On the one hand, there is the notion of perfect adaptation characteristic of Cuvier and the school of Paley, which may be designated conveniently and not too inaccurately as the Bridgewater variety. This includes environmental determinism, that is, the idea of "close" or "strict" adaptation, and also the Cuvierian explanation of structure wholly in terms of "final causes" or "conditions of existence"; it is Bell's "principle of adaptation." According to this view, adaptation – the fit between form and function and between organism and environment – is so close that it can serve as a complete explanation of organic phenomena. This is the assumption of Lyell: Of all the possible ways the creator might choose to adapt an organism to the circumstances under which it is to live, one way is better than any other and is therefore chosen. The structure of the organism is completely dependent on its conditions of existence, and the conditions of existence explain every detail of structure. Conditions of existence may be conceived as broadly as an author wishes. They may include the internal coordination of parts, relations of the organism to the inorganic world, and the interrelations of organisms. But in every case, it is conditions alone that determine structure; that is, there is a strict, an invariant, relationship between particular conditions and particular organic structures: For any given set of conditions there is only one form that is the best possible. If, according to this view, we find a vertebrate animal inhabiting a particular station, its vertebrate structure is not due to a law of unity of type or to heredity. It is due rather to the fact that for that station an animal with a backbone is the best possible. Perfect adaptation in this sense explains even the most fundamental elements of structure ("If there are resemblances between the organs of fishes and those of the other vertebrate classes, it is only insofar as there are resemblances between their functions," Cuvier said).[87]

On the other hand, there is the sort of perfect adaptation that was common to those who rejected the Bridgewater variety. This second variety has room for rudimentary organs, the phenomenon of unity of type, and the fact that apparently there is no strict relationship between environmental conditions and organic form. This might best be called a doctrine of "limited perfection." Its adherents did not believe that organisms are in any sense imperfect, and they certainly did not suppose that they are merely relatively well adapted compared to their competitors. Organisms are perfect, they said; but they are created by laws, and they are only as perfect as is possible within the limits set by the necessity of conforming to these laws. Since the creator's laws are good and well-conceived, adaptation is the general rule. But an animal may possess an organ that serves no function, though it is useful to another animal of the same type. For those who understood perfect adaptation in this way, the basic structure of the previously mentioned vertebrate animal could not be explained by the circumstances in which it lives. Adaptation is not "close" enough for that. Stations that are apparently very similar might be inhabited by animals as different as vertebrates and molluscs. So if this particular animal is a vertebrate, it is because the creator established a law of unity of type in the great classes, or a law of heredity; and part of his plan is that a proportion of all animals will have the vertebrate structure. The animal in question is not then formed solely with reference to conditions. It is formed with reference to the law of unity of type or of heredity, and to the creator's plan. Given its typical or hereditary structure, however, it is as perfect as possible, that is, as well fitted to its environment as it is possible for an animal with its basic structure to be. It is perfect within the limits of the general laws established by the creator.

As far as biological theories are concerned, the main difference between these two varieties of the concept of perfect adaptation is that, according to the first, adaptation is so close that the principle of adaptation, or of the conditions of existence, is a sufficient explanatory principle for organic phenomena; while according to the second, other laws, in addition to adaptation, are required to explain the facts of structure, distribution, and succession. The second clearly gave the biologist, in his efforts to account for these facts, a much wider choice of theoretical possibilities, including even transmutation. For this reason, the distinction is an important one for our understanding not only of Darwin, but of mid-nineteenth-century biology in general.

It is equally important, however, to keep in mind the common ground that these two varieties of perfect adaptation shared, especially in the British setting in which Darwin worked. Both were parts of a view of nature that became scientifically (as well as theologically) orthodox in Great Britain around the end of the seventeenth century and the beginning of the eighteenth. When for various social, political, religious, and philosophical reasons it began to seem desirable to found Christianity on reason (and, at about the same time, to defend the new science as compatible with and useful to Christianity), a premium was placed on arguments that purported to demonstrate from nature the existence and attributes of God. This enforced a view of nature as something which when observed by the light of reason would point unfailingly to a wise, beneficent, and omnipotent creator. All the phenomena of nature were held to be effected by God's laws, which he established in order to produce, in Robert Boyle's phrase, "the ends He proposed to Himself in making the world."[88] To those who accepted this notion, and it proved as attractive to deists as to Christians, nature appeared to be a perfect, harmonious, orderly, well-balanced, stable system, each of whose parts has its purpose (or purposes).[89] After the Restoration, and more especially after

35

the revolutionary events of 1688–9, natural religion and physico-theological arguments were advanced – as by Boyle, John Ray, Newton, and several of the Boyle lecturers – and accepted in part because they were seen as an aid to the maintenance of social and political order against atheists, enthusiasts, and reformers who might upset it.[90] They soon spread widely on the continent, where they served similar purposes for some, while others saw in them primarily a religion suitable for a scientific and enlightened age, a religion all reasonable people could accept.[91] Despite the criticisms of David Hume and others, they retained their force until the middle of the nineteenth century.

After 1789 in Great Britain, natural theology and the view of nature it presupposed and taught acquired an increased ideological importance which it kept throughout the tumultuous years between Waterloo and the Great Exhibition of 1851.[92] This reinforced the idea that nature is a harmonious and orderly system, operating in accordance with purposeful laws established by God. It would probably be fair to say that the strength of this tradition, especially in Britain but also on the continent, meant that for most naturalists educated in the early nineteenth century, whatever their political or religious interests, perfect adaptation had the status not of a postulate of natural theology, nor of an element in a particular ideology, but of a fact apparent to all who took the trouble to observe organisms. For a period after 1789 the extreme teleological interpretation of perfect adaptation, whose antievolutionary implications made it attractive to the guardians of religion, morality, and order, gained prominence, largely through the work of Cuvier.[93] But in the 1830s many naturalists concluded that, contrary to the teleologists' claims, adaptation was not a sufficient explanation of organic phenomena. They did not, however, doubt that adaptation is perfect, because they continued to see all of nature as God's harmonious system.

That the idea of limited perfection is not a deviation from this traditional view is emphasized by the fact that Paley clearly stated the principle that perfection has limits. He noted that the perfection of organisms is not absolute because God, in order to make evident to man his existence, agency, and wisdom, constrained himself to work within the limits imposed by the general laws of matter.[94] While Paley and the biologists and natural theologians of the first third of the century said that perfection is limited only by the laws of matter (or, more commonly, supposed that it is virtually absolute), Darwin, Owen, Carpenter, and others said it is limited also by laws of organized beings, which, like the more general laws of matter, are parts of the harmonious system. The resulting notion of perfect adaptation is quite different from Paley's, but the principle of limited perfection is the same.

Conclusion

In his penetrating examination of the concept of natural selection, Camille Limoges argued that the essential step that allowed Darwin to formulate his theory was his calling into question the traditional conception of the economy of nature. Darwin did this, he said, in the first three transmutation notebooks, and this brought together in their proper relations all the principal elements of the idea of natural selection. When Darwin read Malthus at the end of September 1838, these elements crystallized into the Darwinian theory. Limoges's argument that Darwin at a very early period attacked the traditional conception of the economy of nature can be reduced to two propositions: (1) that the idea of perfect adaptation was the "keystone of the vault" of the traditional conception; and (2) that Darwin denied the perfection of adaptations. Both of these propositions are problematic, as I see it, because Limoges did not distinguish between the two varieties of perfect adaptation. He concluded that Darwin's denial of close adaptation was equivalent to a repudiation of the traditional harmonious view.[95] But in fact, what Darwin was arguing against was teleological explanation, not the idea of harmony. When he abandoned the Bridgewater variety of perfect adaptation, he adopted the notion of limited perfection, which was equally situated within the framework of the traditional conception of nature. Limoges was right to insist on the importance of concepts of adaptation for distinguishing among the views of mid-nineteenth-century biologists, and by doing so he has placed all historians of the period in his debt. But the peculiarly Darwinian concept of adaptation – the relative adaptation that characterizes the *Origin of Species* – is not to be found in the first three transmutation notebooks. What is to be found there is the record of Darwin's continued belief in harmony.

That Darwin was committed to the harmonious conception of

37

nature throughout the pre-Malthus period has been convincingly demonstrated by David Kohn in his important study of the first three notebooks. Kohn's argument – whose incompatibility with Limoges's he points out – is that it was only when Darwin read Malthus that the harmonious conception was shattered for him.[96] Before that time Darwin had seen "imperfect adaptation" as an argument against creationism, but in his theory of transmutation he had continued to suppose that adaptation is very nearly perfect. He revised, rather than rejected, Paley's view of perfect adaptation.[97] My delineation of a clearly defined alternative to the Bridgewater variety of perfect adaptation strengthens Kohn's conclusion, for now there need be no hesitancy about supposing that Darwin rejected the idea of close adaptation and yet continued to believe in the harmony of nature. Others did so too. On the basis of the preceding analysis of the concept of perfect adaptation it is possible to go still further, however, and locate Darwin's pre-Malthus position more precisely. Darwin did not so much revise Paley as adopt a different notion of perfect adaptation. In doing so, he was in good company. This different notion of perfect adaptation was not a unique step along Darwin's path to natural selection, but was rather a viewpoint he shared with many of the leading biologists of his generation.

By looking at Darwin in relation to contemporary biological opinion, we can avoid reading the novelty of his mature theory of natural selection back into his early speculations. We can see more clearly what his rejection of close adaptation was and was not, and we can better understand the perspective from which he viewed the question of transmutation in his early notebooks. In studying Darwin's post-Malthus work the analysis of perfect adaptation is equally useful, for it makes it possible, in tracing the development of Darwin's thought, to see to what extent the structure and development of the theory of natural selection were shaped by ideas about nature that Darwin inherited from the culture in which he worked.

2

Darwin before Malthus

Darwin's belief in harmony and perfect adaptation largely determined the structure of his pre-Malthus theory of transmutation. In the opening pages of the first transmutation notebook, "propagation" of species is presented as an answer to the question of how harmony is maintained during the constant round of geological and climatic changes postulated by Lyell, and throughout all the notebooks Darwin directed his speculations toward explaining the persistence of perfect adaptation despite alterations in the external conditions to which species must conform. Although David Kohn has given an admirable exposition of the main lines of Darwin's pre-Malthus theory, it is necessary to go over some aspects of it again here because it is of crucial importance for understanding the initial shape and the subsequent transformation of the theory of natural selection. When he thought of natural selection, Darwin did not cast aside all the work he had done up to that point. He built his new theory on the foundation he had already laid, and consequently, for many years natural selection in its basic structure closely resembled Darwin's pre-Malthus theory of the maintenance of harmony. Because my primary interest is in natural selection, the emphases of the following study of Darwin's preselection speculations are somewhat different from Kohn's. I especially wish to underscore Darwin's continued belief in perfection and harmony and to indicate how this shaped his attitudes toward variation and adaptation, attitudes that persisted long after Darwin read Malthus. But my account overlaps Kohn's at many points, and, throughout, mine is heavily indebted to his analysis. In particular, I accept without reservation three of his principal theses: that for the whole of the period from July 1837 to September 1838, Darwin (1) conceived of adaptation as an automatic organic response to the environment; (2) conceived of adaptation as absolute, or "well nigh perfect";[1] and (3) had a coherent theory of transmutation in which the process of sexual generation was supposed to produce adaptive variations.[2]

Adaptation by generation

There have been several studies of Darwin's transmutation note-books, most of which have tried to show how Darwin got from his initial belief in transmutation to the theory of natural selection, and some of which have tried to explain on their own terms the explanations of transmutation Darwin employed before he read Malthus.[3] Of these explanations the most frequently noted is Darwin's belief that geographical isolation contributes somehow to the formation of new species. His views on heredity, extinction, and the role of changed habits in producing structural change have also received considerable attention. And one persistent strand of Darwin scholarship has advocated the view that he had a theory of selection, or was looking for one, before he read Malthus.[4] Kohn's study suggests that throughout the pre-Malthus notebooks Darwin adhered fairly consistently to a single – distinct-ly nonselectionist – theory of transmutation to which his various ideas about isolation, extinction, heredity, and so forth are in the nature of accessory hypotheses developed to deal with particular problems or bodies of evidence. The theory, Kohn shows, was fundamentally a theory of generation. Darwin supposed that the process of generation produces not only new individuals, but also new species.

Darwin began his first notebook with a discussion of the "two kinds of generation." There is "the coeval kind," budding, grafting, fission, and the like, which produces offspring exactly like their parent and each other: "all individuals absolutely similar." And there is "the ordinary kind which is a longer process, the new individual passing through several stages (? typical or shortened repetition of what the original molecule has done)."[5] It was the ordinary kind, sexual generation, in which Darwin was chiefly interested. To the question of why it exists, he answered as follows, beginning in a Lyellian vein:

> We *know* world subject to cycle of change, temperature and all circumstances, which influence living beings.
> We see the young of living beings become permanently changed or

subject to variety, according to circumstance, – seeds of plants sown in rich soil, many kinds are produced, though new individuals produced by buds are constant; hence we see generation here seems a means to vary or adaptation. – Again we know, in course of generation even mind and instinct becomes influenced. Child of savage not civilized man. – Birds rendered wild generations acquire ideas ditto. V. Zoonomia. –

There may be unknown difficulty with *full grown* individual with fixed organisation thus being modified, – therefore generation to adapt and alter the race to *changing* world.[6]

Although the mechanisms Darwin imagined to be at work are only vaguely suggested, the general import of the passage is clear. Changes in circumstances somehow influence the generative process, causing the production of young unlike their parents, and the consequence of this is to adapt organisms to changes in external conditions. From this it follows that with changes in conditions, all species have a "tendency to change." It is implied, further, that these organic changes are automatically adaptive, for adaptation is the sole reason for them. This is confirmed a few pages later in the first notebook: "Changes not result of will of animals, but law of adaptation as much as acid and alkali."[7] When the chemist adds a quantity of acid to a quantity of alkali, an automatic change is produced according to the laws of chemical combination. Similarly in nature, when an alteration is introduced into the conditions of life, an automatic change in organisms is produced according to the law of adaptation.

Whatever the law of adaptation might be, Darwin was convinced that it worked through the generative system. Near the end of the second notebook, after nearly a year of thinking about the problem, and only three months before he read Malthus, Darwin restated this fundamental assumption. Why are there two sexes, he asked, when they are not necessary to generation? "I can scarcely doubt final cause is the adaptation of species to circumstances by principles, which I have given."[8] A few passages in the first three notebooks shed some faint light on how Darwin thought about generation and its ability to produce adaptive change. One persistent, though obscure, idea is that this power is connected with the fact that in sexual generation the embryo passes through a series of stages, which Darwin assumed to be a recapitulation of the organism's ancestry: "an originality is given (and power of adaptation is given by *true* generation), through means of every step of progressive increase of organization being imitated in the womb which has been passed through to form that species." How this allows the production of "originality" and adaptation is never specified, but that it does so is asserted as late as September 1838: "the very theory of generation [is] the passing through whole series of forms to acquire differences."[9]

A passage in the second notebook reveals more fully the meaning of the basic proposition with which Darwin began. He believed that the external causes of change – broadly, the "changing world" – affect the adult organism so as to produce an alteration in the organization of its egg or embryo: "Once grant that species and genus may pass into each other, – grant that one instinct to be acquired (if the medullary point in ovum has such organization as to force in one man the developement of a brain capable of producing more glowing imagining or more profound reasoning than other, if this be granted!!) & whole fabric totters & falls."[10] Again, it is not explained why this change in the ovum always produces a suitable, an adaptive, alteration in the offspring. One further passage, however, displays another facet of Darwin's thinking about the generative process, and this perhaps furnishes a clue to how "true generation" gives the power of adaptation. "There is probably [a] law of nature," he said, "that any organ which is not used is absorbed. – this law acting against hereditary tendency causes abortive organs. – The origin of this law is part of the reproductive system, – of that knowledge of the part of what is good for the whole."[11] If this sort of thinking were applied to the problem of adaptation – and this would be consistent with the tenor of all Darwin's statements on generation in the first three notebooks – it would lead to a proposition something like the following: External conditions and the habits of the organism affect the reproductive system and give it "knowledge" (presumably Darwin did not mean this literally) of what the offspring's needs will be; this causes the production of appropriate variations from the form of the parent. At one point, though only briefly, Darwin wondered whether the effect of conditions on the reproductive system might be through the medium of the parent's mind. If so, his problem would apparently be solved: "Can the wishing of the Parent produce any character in offspring? Does the mind produce any change in offspring? If so, adaptation of species by *generation* explained?"[12]

Despite the imprecision of his ideas on how generation produces adaptation, throughout the pre-Malthus period Darwin remained convinced of the validity of the two central propositions of his theory, that organic change is accommodation to circumstances and that this accommodation is effected primarily through sexual generation. Adaptation is the final cause of generation. "Why does individual die," that is, why are there generations? "To perpetuate certain peculiarities (therefore adaptation), and to obliterate accidental varieties, and to accommodate itself to change *(for, of course change even in varieties is accommodation)*."[13] In the theory of natural selection variations are generated on which selection works to produce adapted forms. But in this earlier theory, the variations are themselves adaptive. All change is adaptation. Darwin's discussion of monsters offers perhaps the most striking evidence for this. He suggested that monsters differ from adaptive novelties only in that they are not adapted for the whole life of the organism. "Even a deformity may be looked at as the best attempt of nature under certain very unfavoured conditions, – as an adaptation, but adaptation during earliest existence; if whole life then real adaptation."[14] Others have cited this passage to show that Darwin believed adaptation was "imperfect." If even deformities are "adaptations," then in no sense can adaptations be thought of as perfect.[15] It seems to me, on the contrary, to show that Darwin imagined a deterministic law of adaptation that invariably, even under "unfavoured conditions," works to produce a fit between circumstances and organic form.

For Darwin "adaptation of species by *generation*" was a general expression for one part of the grand system of laws that he supposed produced all the effects of the organic and inorganic worlds and maintained the harmony between them. But though adaptation was in his view the final cause of the propagation of new species, he did not at first believe that explaining exactly how adaptation is produced was his most pressing task. The law of adaptation could be assumed for the time being, its existence guaranteed by the facts of nature, as he and his contemporaries perceived them, and by his conviction that the whole system of laws was the work of a benevolent creator. Of more fundamental importance to his enterprise was working out the implications of the laws of generation and heredity. This was necessary in order to establish the probability, or at least the possibility, of transmutation, which Darwin treated as a prerequisite for finding the law of adaptation.[16] Moreover, the laws of generation at first appeared to be more accessible to investigation. Accordingly, it was to them that Darwin most consistently attended in the transmutation notebooks. Four of these played particularly important parts in his pre-Malthus theory: blending inheritance, or the "law of intermarriages"; the "law of hybrids," including the tendencies to infertility and to reversion to parental forms; the hereditary fixing of characters that remain "long in blood"; and the loss of desire when strains are too closely interbred. I will take up each in turn, for Darwin's discussions of these laws reveal particularly well some of his most firmly held and long-persistent beliefs about the connection between variation and adaptation.

The belief that in offspring produced by the sexual union of two individuals there is a blending of the characters of the parents, so that the young are intermediate between them, was commonplace in Darwin's day and was accepted by him as a generally valid rule.[17] It is introduced in the opening pages of the first notebook and assigned a function of central importance in Darwin's theory. Any theory of change in the organic world must be able to account for the universally observed fact that species appear to be more or less constant at every place and time. The "tendency to vary by generation" that Darwin proposed was potentially in conflict with this observation. To resolve this difficulty Darwin invoked the "beautiful law of intermarriages partaking of characters of both parents." Taken by itself the tendency to vary might be expected to cause species to be fragmented into local subsections that have been adapted to different circumstances, or, as he suggested at one point, into independent individuals. But blending assures that these local and individual differences will be merged to give the species a uniform character.[18]

Blending has a further major function in Darwin's theory. It promotes adaptation to change by preventing slight local fluctuations from interfering with the process. "One of the final causes of sexes [is] to obliterate differences," Darwin said. "Final cause of this because the great changes of nature are slow. If animals became adapted to every minute change, they would not be fitted to the slow great changes really in progress."[19] This passage, written in September 1838, provides a useful insight into Darwin's theory just before he read Malthus. The implication is clear that all external change, however slight or transitory it may be, produces adaptive responses in organisms. But the more trifling changes are blended out, so that only those that are needed to adapt the species to the "great changes really in progress" become a permanent part of its hereditary character.

Blending also presents problems, however. If a change in circumstances affects the entire range of a species, then the whole species will become adapted to the new conditions. But if one part of the range is persistently subjected to environmental change, the members of the species that live there may be unable to adapt because changes they undergo will tend to be blended and made intermediate by crossing with unaltered neighbors. "Complete adaptation" will thus be prevented.[20] Early on Darwin suggested that geographical isolation, especially on oceanic islands, would promote change.[21] But it was necessary further to show that after a period of isolation the accumulated change would not be lost by blending with the unaltered parent form. For his theory of transmutation to work there must be found laws that tend to preserve change after it has occurred. This was a problem to which Darwin devoted a great deal of thought and for the solution of which he collected from breeders, naturalists, and travelers numerous facts and generalizations relating to crossing, hybrids, and so forth.[22] He noted that hybrids tend to be infertile; that they tend to resemble one parent more than the other; that when fertile offspring are produced by crossing, they tend in subsequent generations to revert to the parent types; that among forms sufficiently unlike, there is a "repugnance to intermarriage"; and that if they are still more different, crossing is physically impossible. He thought of these situations as forming a continuum.[23] At one extreme is the case in which two organisms are only slightly different. Blending prevails, and the offspring are truly intermediate. When the parents are more unlike, the offspring resemble one more than the other: "Probably this is first step in dislike to union, offspring not well intermediate." Lessened fertility and reversion are the next steps, followed by "dislike to union," complete infertility ("no offspring"), and "physical impossibility to marriage."[24] Which stage was crucial in the formation of new, permanent species Darwin did not specify. At one point he suggested "repugnance" was required: "A species as soon as once formed by separation or change in part of country, repugnance to

45

intermarriage – settles it." Elsewhere he urged that the fact of reversion effectively established the possibility of transmutation: "Have *races* of Plants ever been crossed really; if there is any difficulty in such marriages or offspring show tendency to go back – there is an end to species."[25] The significance of the whole line of speculation and the connection between the laws of blending and hybrids are nicely expressed in the following formulation: "If after isolation (seed blown into desert or separation of mountain chains &c.) the species have not been *much* altered they will cross . . . & so [they] make that sudden step species or not."[26] A sufficient amount of change must be produced or the laws of hybridity will be unable to preserve it from obliteration by blending.

A third law, one that became important enough to Darwin that he sometimes called it "my theory of generation," is that the longer a character is possessed by a race of organisms – or, the greater the number of generations through which it is preserved – the more firmly fixed it becomes in the race's hereditary make-up.[27] Darwin derived this law from the zoologist William Yarrell's statement that when older races are crossed with younger, the hybrids more closely resemble the older; the "oldest variety impresses the offspring most forcibly."[28] As reformulated by Darwin, the law is "what has long been in blood, will remain in blood, – converse, what has not been, will not remain."[29] He believed that this would explain a number of phenomena. It served first of all to account for the fact of hereditary resemblance, which in a sense is called into question by the introduction of a principle of change, a "tendency to vary by generation." If there is such a tendency, why does it happen that in most respects children resemble parents and grandparents; why might not the variation be in such a fundamental character as the basic skeletal structure, for instance? But if characters become more "impressed in blood" with time, this is explained, for some, such as basic structure, will then be so firmly fixed as to be virtually invariable: "If varieties produced by slow causes, without picking become more & more impressed in blood with time, then generation will only produce an offspring capable of producing such as itself."[30] In similar fashion, the characters that are constant throughout large groups of organisms, the typical structures on which classification depends, are explained, as is the law of succession – that in a given region the modern animals are the same type as the extinct. Those characters that are more firmly fixed will be constant, or only slightly variable. "Where any structure is general in all species in group we may suppose it is oldest, & therefore least subject to variation."[31] Darwin thought that frequently these would be internal parts, because external influences first affect external form, and consequently internal parts will have varied least and be more fixed in their character.[32] In a different context, fixity of character was invoked to explain extinction. A species that has remained adapted to the same circumstances for many generations may have so rigid a hereditary constitution that it is unable to adapt to a rapid change of conditions and so perishes.[33]

By far the most important function hereditary fixing serves in Darwin's early theory is to promote transmutation. It does this by causing the principal phenomena of crossing – reversion to parent forms, infertility, and so forth – that preserve change once it has occurred. In the second notebook Darwin explained that because a "variety when long in blood gets stronger & stronger" it may at best be possible, when crossing such varieties, to produce one offspring that is unlike its parents – that is, a hybrid. But to produce a whole succession of generations that are unlike the parents "would go against the [hereditary] tendency." Consequently, the offspring of hybrids tend to revert to the form of their grandparents. But if the hybrid is too unlike its parents, this is impossible; instead of producing offspring that "go back to grandfather," the hybrid is infertile. "It is not difficult to see," Darwin said, "that it is less repugnant to nature to produce one offspring unlike itself, than to produce that [one] capable of producing itself alike. – in one case it changes one, in other it changes thousands in futurity."[34] Elsewhere Darwin said that hybrids tend to be infertile or have offspring that "go back" because, in effect, they have no hereditary tendency of their own. Their organization has not been "long in blood," but has been acquired in a single generation by the mixture of two very different stocks.[35] The implication of the law, when stated in this form, is that the hereditary tendency is composed only of changes that have been acquired slowly and in small increments. Rapid and large changes are short-lived, for the offspring of organisms so changed are infertile or revert to "grandfather."[36] The small and slow changes that result from slowly changing conditions, however, become "congenital" and are fixed in the blood; "by a

47

succession of generations, these small changes become multiplied, & great change [is] effected."[37] That is, once the change is great enough, the hereditary tendency prevents its being obliterated by crossing.

By the middle of the second notebook, Darwin saw hereditary fixing as the most important of the laws of generation. Not only did it help account for a host of biological phenomena – including hereditary resemblance, the existence of typical structures in large groups, the law of succession – but it seemed to explain how change is accumulated and how it is preserved.

The law of hereditary fixing was largely Darwin's own creation, elaborated expressly to support his theory of transmutation. By contrast, the fourth law, that inbreeding causes a loss of desire (as well as a decrease in fertility), was a maxim that many breeders accepted and which Darwin therefore had to integrate into his system.[38] This proved to be difficult, because he could not at first see what its importance could be. It seemed, indeed, to stand in the way of change in small populations, such as those on islands, where a certain amount of inbreeding is inevitable. But by considering the converse of the law – that when the parents are somewhat different, they will tend to produce more fertile offspring – he was able to discover two functions it might serve. The effect of the law and its converse is to discourage inbreeding and encourage outbreeding. This promotes blending and consequently helps establish the uniform character of species: "The increased fertility of slightly different species and [the] intermediate character of offsprings accounts for *uniformity* of species."[39] In other words, the advantage of intermarrying, together with the blending that results from it, explains uniformity. The fact that blending is thus encouraged helps explain also why there is no gradation of form where there is a gradation in conditions, a fact that Darwin feared might be used as an argument against transmutation. The "law of small differences producing more fertile offspring" encourages cross breeding, and this "prevents perfect change."[40]

The second function of this law touches the very heart of Darwin's theory of transmutation. The final cause of sexual generation for Darwin is the production of change. The passions are nature's way of promoting sexual generation, and hence of promoting change. And the failure of the passions is its way of discouraging inbreeding, which would prevent the production of change. Why inbreeding does not produce change Darwin explained as follows: when two very similar individuals unite, their offspring will exactly resemble both of them, for there are no differences to blend. Instead of producing change, inbreeding, like asexual generation, makes all individuals similar:

> Generation being means to propagate & perpetuate differences, (of body, mind & constitution) in the end frustrated, when near relations & therefore those very close [i.e., similar] are bred into each other . . . Every individual foetus would reproduce its kind was it not for the necessity of some change. Without some small change in form, ideosyncrasy or disposition were added or subtracted at each or in *several* generations, the process would be similar to *budding* which is not object of generation. – therefore passions fail . . . The upshot of all this is that effect of male is to impress some difference to make the *bud* of woman not a bud in every respect . . . The very theory of generation being the passing through whole series of forms to acquire differences, if none are added, object failed, & then by that corelation of structure, desire fails.[41]

If the loss of desire is to be avoided, differences must constantly be added. The differences, Darwin said, may be caused by external circumstances affecting an individual during its own life, or they may be due to differences in its parents, that is, to the effects of external circumstances on previous generations.[42] Here in his discussion of how differences are constantly added we encounter once again Darwin's belief that all organic change is an adaptive response to changes, however slight, in external conditions. The circumstances that cause change in organisms "must be external," he said. If the changes in external circumstances are "always of one nature," a new species is formed. But if they are mere fluctuations, then organic changes fluctuate with them; they "oscillate backwards & forwards & are individual differences (hence every individual is different)."[43] It is these "individual differences" that are constantly being blended and prevented from accumulating, so that there will only be adaptation to the "great changes really in progress." Yet even these differences are adaptive, that is, appropriate responses to slight external changes.

By September 1838 Darwin was in possession of a set of principles that he thought explained how sexual generation produces change and how change, once produced, is preserved. Through the medium of the reproductive system, changes in external conditions call forth corresponding changes in organisms. Even the slight changes that are always occurring, such as fluctuations in the severity of winter or the amount of rainfall or the local movements of animals, cause slight differences in constitution, structure, mind, and so forth. These slight differences are normally blended out or reversed by further fluctuations in conditions. Organic changes are also produced in response to great slow changes of conditions, whose causes are fundamentally geological, and these changes may be preserved. (Darwin's law of hereditary fixing, according to which some structures are more variable than others, implies that slight fluctuations in conditions will chiefly affect external parts, while greater changes may produce alterations in the more firmly fixed, usually internal parts of organisms as well.) If the great external changes affect individuals in only a part of the continuous range of a species, crossing with unaltered forms may prevent complete adaptation. But if such changes affect individuals of a species over all of its continuous range, as on a continent with no barriers, or if they affect some isolated portion of a species, as on an island, then organic changes will accumulate. Alterations that persist through many generations become more and more firmly fixed in the hereditary constitution. Consequently, if a new species is brought back into contact with its unaltered parent species, the two will not readily cross, and the offspring of crosses will tend to be infertile. Blending thus will not obliterate the changes that have occurred. On the other hand, if the change is not great enough or continued long enough, the intermarriages will produce fertile, intermediate offspring, and the alterations will be completely or partially lost.

Although his laws of generation could not by themselves account for adaptation, Darwin in developing and applying them never wavered in his conviction that their primary purpose was to adapt organisms to a changing world. They served other purposes, too, such as explaining the uniformity of species, the sterility of hybrids, reversion, and the existence of the taxonomist's natural groups. But the reason for their existence in the creator's system was to assure that when external conditions change, as geology shows they must, the fit between organic and inorganic worlds is not destroyed. Consistently, Darwin assumed that all organic change, whether it produces monsters or "individual differences," is "accommodation." By showing how such changes could be accumulated and preserved, his laws of generation provided a seemingly secure foundation for his transmutationism.

Adaptation and extinction

Two other features of Darwin's pre-Malthus theory demand consideration here, partly because they represent his solutions to important problems, but even more because they show some of the ways in which Darwin's thinking was governed by assumptions of harmony and perfection. The explanation of adaptation that Darwin adopted at this time involved none of what we normally consider to be characteristically Darwinian conceptions, such as the idea of relative adaptation, the idea of chance variation and differential fitness of variant forms, and the idea of struggle. The same is true of his principal explanations of extinction.

Although he did not initially think so, by the spring of 1838 Darwin apparently believed that a successful theory of transmutation would have to explain adaptation. Toward the end of his first notebook he wrote, "If my theory true, . . . we are led to endeavour to discover *causes* of changes, – the manner of adaptation . . . becomes full of speculation and line of observation."[44] The "theory," it is implied, can alone serve to reorganize and redirect the study of natural history. Discovery of the law of adaptation will then be one result to be hoped for after transmutation is established as the basis for this redirection. Up to this point Darwin's speculations on adaptation had been desultory and had led no further than the guess that "wish of parents??" might explain it. But from the evidence of the second notebook it seems that he subsequently devoted considerably more thought to the problem, with the result that he soon found a better theory, one that he adhered to until he read Malthus. He argued that frequent repetition of habitual and instinctive behavior, in response to altered environmental conditions, causes changes in structure. This is first stated shortly as "instinct goes before structure." A hypothetical example of the effects of habit is soon added: "fish being excessively abundant & tempting the Jaguar to use its feet much in swimming, & every developement giving greater vigour to the parent tending so ![to?] produce effect on offspring." By the middle of the second notebook it has become "of the utmost importance to show that habits sometimes go before structure," and not long after, an example is found in which "we SEE structure gained by habit." The most extensive statement of the habit-instinct-structure theory is the following:

> Reflect much over my view of particular instinct being memory transmitted without consciousness, a most possible thing see man walking in sleep. – an action becomes habitual is probably first stage, & an habitual action implies want of consciousness & will & therefore may be called instinctive. – But why do some actions become hereditary & instinctive & not others. – We even see they must be done often to be habitual or of great importance to cause long memory, – structure is only gained slowly. Therefore it can only be those actions which *many* successive generations are impelled to do in same way . . . Memory springing up after long intervals of forgetfulness, – after sleep strong analogies with memory in off-spring, or simply structure in brain people & senses recollecting things utterly forgotten – Some association in such cases recall the idea it is scarcely more wonderful that it should be remembered in next generation . . . Analogy a bird can swim without being web footed yet with much practice & led on by circumstance it becomes web footed. Now man by effort of memory can remember how to swim after having once learnt, & if that was a regular contingency the brain would become web-footed & there would be no act of memory.

This, epitomized, becomes: "according to my views, habits give structure, [therefore] habits precede structure, [therefore] habitual instincts precede structure."[45]

This, of course, is Lamarck's doctrine of use and disuse in response to environmentally induced needs.[46] Darwin merely revised it slightly to allow its easy incorporation into his theory of generation. Because Lamarck supposed that organisms may alter in the course of a single lifetime, his version of the habit-structure theory required the generative process only to transmit acquired differences. Had Darwin followed exactly the same line, we would have to conclude that his quest for the "manner of adaptation" had led him rather far from the idea of "adaptation of species by *generation*." But since he continued to refer to his entire theory of transmutation as "my method of breeding," it appears that this is not so.[47] Darwin, like Lamarck, thought that structural and other changes might occur in the lifetime of a single individual and that such changes would tend to be inherited. But he seems to have had in mind more than this. By "habit goes before structure" he meant, I think, that habitual actions performed during a number of generations will gradually produce structural change in succeeding generations. In other words, it is new habits, formed in response to environmental change, that convey to the reproductive system "knowledge" of what differences in structure the succeeding generations require. All of the passages in the notebooks that bear on this (including those quoted just above) are ambiguous, and so I do not wish to insist on it very strongly. The advantage of interpreting Darwin's Lamarckism in this way is that it is consistent with the primacy his theory of generation continued to have in his speculations even after he had become a Lamarckian. At the end of the third notebook, as at the beginning of the first, Darwin held that the function of the generative system is to produce, as well as transmit, differences.[48]

Darwin's decision in 1838 to explain adaptation after the fashion of Lamarck was in keeping with his general approach to biological problems in the early notebooks, an approach that in many ways was much like Lamarck's. Lamarck was a deist convinced of the rationality and harmony of the natural world. At least one of the motives that led him to his theory was the desire to avoid the disruption and irrationality that he thought was implied by the idea of extinction.[49] Darwin too viewed the world as a harmonious system and was engaged in an effort to explain how the organic world continues to be adapted to the inorganic. His two answers – his theory of generation and his Lamarckian explanation of adaptation – both treat adaptation as an essentially automatic response of species to environmental change. Changing circumstances lead to new habits and to the production of new, adaptive variations by the generative system. Thus the laws of nature effect the creator's purpose that the harmony of the world persist amidst changes.

The assumption that adaptation is automatic was natural for one with Darwin's outlook. But it was not without its drawbacks. If adaptation is automatic, why do species become extinct? The problem arose for Darwin because of his position as a transmutationist who believed in harmony and at the same time admitted the fact of extinction. After the work of Cuvier, and the blossoming of paleontology into a recognized branch of anatomical studies, virtually all naturalists had had to incorporate extinction into their conceptions of the economy of nature. By 1830, one such as Lyell could explain it almost effortlessly. When a change in external conditions results in a species' being no longer well adapted, it becomes extinct. In Darwin's phrase, "non-adaptation of circumstances" is the cause. In outward form, Darwin's view of nature and its changes was very like the view of Lyell and the great majority of naturalists in his day, who believed in an established fitness of things. They supposed that at any one time there are in the world a certain number of stations, and these are inhabited by plants and animals that are well adapted to them. When geological and other changes occur, some organisms are rendered unfit. New, well-adapted forms are then produced to replace them, and the harmony thereby is maintained. But whereas others, who did not believe in transmutation, said that the unfit forms simply become extinct, Darwin said that they produce offspring that are well adapted, so that their line, at least, continues, rather than perishes. In constructing a theory to preserve harmony, Darwin, like Lamarck, almost made impossible such a disharmonious event as extinction. And while this had been Lamarck's intention, it was not Darwin's, for by the 1830s extinction was no longer in doubt, and Darwin was especially well aware of it from his own experience as a fossil collector in South America. Furthermore, one specific aspect of Darwin's conception of the harmony of nature seemed to make extinction of whole lineages an inevitable part of the process of organic change.

Numerous times in the notebooks Darwin expressed his belief that the number of species must be more or less constant over geological time. As others have pointed out, this was an idea he owed to Lyell. In some of his geological notes from the *Beagle* voyage he wrote:

> If the existence of species is allowed, each according to its kind, we must suppose deaths to follow at different epochs, & then successive births must repeople the globe or the number of its inhabitants has varied exceedingly at different periods. – A supposition in contradiction to the fitness, which the Author of Nature has now established.[50]

There was in the 1830s no necessary connection between the harmonious view and the idea of constant population. A geologist who supposed the earth's history to have been directional could conceivably have believed in a "fitness" consisting of either an increasing or a declining or a constant number of forms. However, the only geologist who by 1835 had seriously attempted a synthesis of biology and geology was Lyell, and his synthesis tied the organic world to a cyclically fluctuating inorganic world.[51] In Lyell's "fitness," population might fluctuate with the great cycle of climate, or it might remain constant, as did the relative proportions of land and sea according to his theory, but it would not vary exceedingly. This view of fitness was Lyell's idiosyncratic expression of a traditional attitude toward the economy of nature, which he shared with virtually all of his contemporaries, teleologists and antiteleologists alike: that organisms are adapted chiefly to physical conditions, especially climate, rather than to other organisms; that the number of places or stations in the world is determined largely by physical conditions; and that the successive appearance of new forms of life during geological time has occurred as a series of responses to changes in physical conditions. This, in its Lyellian form, was Darwin's attitude too, even after he became a transmutationist; hence his long-continued belief in the numerical constancy of species.[52]

Darwin's discussions of constancy are closely bound up with his discussions of one of the principal features of his evolutionary system – branching, the multiplication of species. From a very early stage in his speculations Darwin assumed that evolution is a branching, rather than a linear, process. Isolation and changes in conditions multiply species, for one parent may generate several offspring. But multiplication, if unchecked, would lead to an increase in population, which Darwin's conception of harmony would not admit. His belief in branching therefore required belief in extinction. Darwin noted this consequence the first time he invoked the image of the tree of life:

> With this tendency to change (and to multiplication when isolated) requires deaths of species to keep numbers of forms equable. But is there any reason for supposing number of forms equable: This being due to subdivisions and amount of differences, so forms would be about equally numerous . . . Organized beings represent a tree, *irregularly branched*; some branches far more branched, – hence genera. – As many terminal buds dying, as new ones generated.

A few pages later he wrote, "I think Case must be that one generation then should have as many living as now. To do this and to have many species in same genus (as is) *requires* extinction."[53] Extinction is necessary, but what is its cause? The nature of Darwin's theory, together with the peculiar features of the South American extinctions of large Mammalia, made this a very difficult question to answer.

One purpose of Lamarck's theory had been to offer a means of accounting for apparent extinctions: Species have changed their forms, he said, so that we no longer recognize them in the fossils we discover.[54] A modified version of this explanation was a natural part of Darwin's theory also. The "Fathers" that are no longer well adapted after a change in conditions generate new species, while they themselves become extinct. But extinction of the "Fathers" was all that Darwin's theory could accommodate easily, and it was insufficient, with respect both to the evidence and to the branching-constancy antithesis. The fossil evidence, especially the large South American mammals Darwin had found, indicated not only that the parents of modern forms were extinct, but that there had been an "absolute end" of some forms; that is, some had not propagated new species.[55] Why had they not? Why had the automatic process of adaptation failed to work? Even if Darwin had succeeded in imagining that *Toxodon* and *Megatherium* had left descendants and therefore fell into the explainable category of extinct "Fathers," this would not have solved his problem. The extinction of the "Fathers" alone would be insufficient given Darwin's belief in constancy. Branching means that one "Father" gives rise to many offspring. If the population is to remain constant, not only must the one "Father" become extinct, but several of his siblings must also, and without leaving any progeny: "if each species as ancient (I) is capable of making 13 recent forms. Twelve of the contemporaries must have left no offspring at all, so as to keep number of species constant."[56]

In order to explain the extinction of forms that left no descendants, Darwin had to imagine conditions under which his automatic mechanism would fail to produce new adaptations. Eventually, consideration of the laws of generation suggested two types of situations in which this might occur. In one, crossing with unaltered forms would cause extinction; in the other, Lyellian local catastrophes. Twice in the first notebook Darwin argued that in some cases the effect of blending inheritance might be to produce immutable, or nearly immutable, species: "Those species which have long remained are those – ?Lyell? – which have wide range and therefore cross and keep similar. But this is difficulty: this immutability of some species."[57] There is no indication, however, that Darwin yet saw this as an answer to the extinction problem. But in the second notebook, in the midst of a summary of what his theory could explain, Darwin used crossing to account for extinction: "Animals having wide range, by preventing adaptation owing to crossing with unseasoned people would cause destruction."[58]

Further on in the second notebook, rapid geological change is called on for the same purposes. This might seem inconsistent in such a Lyellian geologist as Darwin, but in fact Lyell himself had set the precedent, invoking local catastrophes to forestall Lamarck's gradual process of transmutation. Lyell believed that the causes of geological change are slow and uniform. But slowly acting causes, he said, sometimes produce very rapid changes, as when the gradual erosion of a natural dam results in a sudden and perhaps very large and destructive flood.[59] This was an idea Darwin found useful:

> Changes in structure being necessarily excessively slow they become firmly embedded in the constitution, which [is an-] other marked difference in the varieties made by . . . nature & man. – The constitution being hereditary & fixed, certain physical changes at last become unfit, the animal cannot change quick enough & perishes. – Lyell has shown such Physical changes will be unequally rapid with respect to their effects.[60]

This appears to have been Darwin's main explanation of extinction (other than of "Fathers") right up to the time he read Malthus, for as late as September 7, 1838, he reiterated it.[61]

Darwin's solutions to the extinction problem are interesting for what they add to the picture of his faith in adaptation. Species become extinct not because they are inherently inferior or constitutionally incapable of producing new species. The fitness of nature requires extinction, but it occurs not because the potentially burgeoning number of species that results from branching comes up against the requirement of constancy and less well adapted forms perish in the ensuing conflict. For Darwin, there are no less well adapted forms, except when conditions have changed; and even then, all possess the generative mechanism that produces new adaptations. For extinction to occur, that mechanism must be stymied by peculiar external conditions, either catastrophic change or the absence of isolating barriers to prevent crossing. In this, as in all his other theoretical work before the fall of 1838, Darwin's speculations appear to have been guided and informed by the related ideas of harmony and perfect adaptation.

Conclusion

According to Darwin's pre-Malthus theory, the generative system, in response to changing external conditions, produces variations that are adapted to the altered circumstances. The cause of organic change is always assumed to be external, while the change itself is produced by and expressed through the reproductive system. The idea that variation is dependent on external change is reinforced by the law of hereditary fixing of characters, which implies that when the hereditary tendency is not disturbed to any appreciable degree, offspring will resemble their parents, within the narrow limits imposed by the individual differences between the parents (and these individual differences are themselves ultimately the result of slight variations in conditions). It is assumed further that all variation is accommodation to change.

In this view of variation, Darwin deviated very little from the opinions of his contemporaries. This was to be expected, I think, for it was from his contemporaries that he learned about these things. Through the medium of Lyell's *Principles of Geology*, in particular, Darwin was acquainted at an early period with the ideas of a large number of leading biologists; and, most helpfully, Lyell explained the bearing of their work on the question of transmutation. Cuvier, on whom Lyell depended heavily, had said that heat, the quantity and type of nourishment available, and other unspecified causes influence the development of organisms and, within the limits of the specific character, produce varieties.[62] Lyell, drawing also on Lamarck, Blumenbach, and others, gave a more elaborate statement. Naturalists do not assume, he said, that the organizations of plants and animals remain absolutely constant and invariable. All are aware that circumstances influence habits and habits may alter the structure of an organism. In some species a considerable degree of variability, in others, less, may be required in order to enable them to "accommodate themselves, to a certain extent, to a change of external circumstances." According to Lyell, changes in conditions produce variations in organisms, and variations are "accommodations" to these changes.[63] This was Darwin's view also.[64] His theory differed from Lyell's only in that Lyell said the power of accommodation of each species is strictly limited by its original specific character, while Darwin said that accommodation can continue indefinitely, as long as there are changing conditions to require it. The difference is significant, but so is the similarity. The basic structure of Darwin's theory followed directly from his traditional understanding of variation and adaptation. It is a theory of organic response to external change. Transmutation is not a constant, but rather an intermittent, process. It is not continuously tending to produce new and better adapted forms. In the absence of external change, all forms are well adapted; only when changes occur do some become not well adapted. The purpose of Darwin's mechanism for transmutation is to adjust organisms to such changes in conditions, and the essential element of the mechanism is a generative system that produces variations when external conditions affect it. The source of Darwin's attitude, as well as Lyell's, toward the production of variations was the harmonious view of nature, in which organisms were presumed to be always perfectly adapted to conditions.

Darwin's pre-Malthus theory might best be described as the unconventional outcome of the conventional project of attempting to discover the fixed laws that constitute the harmonious system of creation. Not only did Darwin not reject the harmonious view. It seems fair to conclude that it had not occurred to him to think of nature in any other way.[65] He continued to see nature as the culture of early-nineteenth-century Britain taught him to see it,[66] and this should not be surprising. What motives were there for Darwin to adopt an alternate perspective; and what sources were readily available to him to suggest one? The idea of transmutation by itself certainly did not require any departure from the traditional conception. An alternate perspective, insofar as Darwin ever arrived at one, appears only to have been gained gradually and piecemeal in the months and years after he read Malthus and formulated the theory of natural selection.

3

Natural selection and perfect adaptation, 1838–1844

Near the end of September 1838, Darwin read Malthus's *Essay on the Principle of Population* and hit upon the leading idea of a new theory. He saw in population's geometrical power of increase "a force like a hundred thousand wedges trying [to] force every kind of adapted structure into the gaps in the œconomy of nature, or rather forming gaps by thrusting out weaker ones."[1] Since the discovery of the notebook pages containing this sentence, it has been clear beyond any reasonable doubt that Darwin's reading of Malthus was a crucial step in the formation of the theory of natural selection. [2] The question of exactly what Darwin got from Malthus has generated a sizable body of scholarship, but fortunately it is not necessary to review it here.[3] It is enough for my present purposes to say that Malthus led Darwin to see the power of reproduction in a new light, as a potent force which by its wedging action produces adaptation. The question I want to pose instead is this: Given the fact that Darwin read Malthus and as a result got hold of a new theory, to what extent did this cause him to revise his views on harmony and adaptation; and, more importantly, to what extent did these views shape his new theory? The answer I will suggest is that while his reading of Malthus gave Darwin a new theory of organic change and adaptation, it did not immediately alter his conception of nature, nor was his new theory free from the effects of his old assumptions. This seems to me in itself to be a reasonable proposition. While it is easy to imagine that a person's sudden insight may produce a new hypothesis, a new way of solving a problem, I find it more difficult to believe that such an insight can with equal speed alter one's entire world view, from one of harmony to one of discord, for instance. It is only in reflecting on and working with a new idea that its wider implications begin to appear. In many, probably most, instances, all of the consequences of a revolutionary proposal are never recognized by the person with whom it originates. Such conjectures aside, the evidence in Darwin's case seems to me persuasive. The new theory required some little time to effect a major revision in his conception of nature and a considerably longer time to eradicate some of the assumptions associated with the traditional viewpoint. When he read Malthus, Darwin did not at once abandon his belief in perfection and harmony. Although he would later do so, he did not immediately view the organic world as a scene of struggle and discord, productive only of relatively, that is, less-than-perfectly, adapted creatures. Instead, he initially pictured natural selection as operating in precisely the same natural theological context in which Malthus had set his principle of population – the context of a system of beneficent laws designed to produce certain preordained ends. Within a few weeks he gave up the idea that every effect of the creator's laws was included in the original design. But for a long time he continued to believe in perfect adaptation, and this determined the structure of his theory in its early years. As late as 1844 that structure was essentially the same as in the opening pages of his first transmutation notebook, written when Darwin still fully shared traditional beliefs about the harmony of the creation.

The impact of Malthus

Let me begin by setting down Darwin's first record of his new insight:

> [Sept.] 28th We ought to be far from wondering of changes in numbers of species, from small changes in nature of locality. Even the energetic language of Decandolle does not convey the warring of the species as inference from Malthus – increase of brutes must be prevented solely by positive checks, excepting that famine may stop desire – in nature production does not increase, whilst no check prevail, but the positive check of famine and consequently death. I do not doubt everyone till he thinks deeply has assumed that increase of animals exactly proportionate to the number that can live – Population is increase at geometrical ratio in FAR SHORTER time than 25 years – yet until the one sentence of Malthus no one clearly perceived the great check amongst men – there is spring, like food used for other purposes as wheat for making brandy – Even a *few* years plenty, makes population in man increase & an *ordinary* crop causes a dearth. take Europe on an average every species must have

same number killed year with year by hawks, by cold &c – even one species of hawk decreasing in number must affect instantaneously all the rest – The final cause of all this wedging, must be to sort out proper structure, & adapt it to changes – to do that for form, which Malthus shows is the final effect (by means however of volition) of this populousness on the energy of man. One may say there is a force like a hundred thousand wedges trying [to] force every kind of adapted structure into the gaps in the œconomy of nature, or rather forming gaps by thrusting out weaker ones. – [4]

From our post-1859 vantage point these passages are of great interest because they are so obviously the basis for the theory of natural selection. But the very obviousness of this fact may pose problems of interpretation. We know so well what natural selection is that it requires some care not simply to read that knowledge into Darwin's notes. Natural selection, we know, is based on struggle and merely relative adaptation, instead of harmony and perfect adaptation. Did Darwin see that this was so when he first read Malthus? To avoid the danger of finding just what we expect in these passages, we need to approach them not from the perspective of the *Origin of Species,* but rather from the perspective of Darwin's speculations in the period leading up to the reading of Malthus.

Viewed in this light, one sentence strikes me as particularly noteworthy: "The final cause of all this wedging, must be to sort out proper structure, & adapt it to changes. – *to do that for form, which Malthus shows is the final effect . . . of this populousness on the energy of man.*" This points directly to the hopeful or harmonious interpretation of the principle of population that was common to Malthus and to many of the natural theologians of Darwin's day. Malthus is remembered chiefly for his pessimism. But it should not be forgotten that the limited prospects he saw for improvement of the condition of the "lower classes" were in his opinion a consequence of the established order of creation. Like most of the political economists of his school, Malthus supposed that society operates according to divinely appointed laws, which laws must therefore be on the whole good. Writing in the wake of the French Revolution, he was concerned not merely to explain the operation of his law of population, but also to show how it fit into the creator's system. By demonstrating "how little the price of labour and the means of supporting a family depend upon a revolution," he hoped to induce "every man in the lower classes of society . . . to bear the distresses in which he might be involved with more patience, . . . feel less discontent and irritation at the government and the higher classes of society on account of his poverty, . . . become more peaceable and orderly, . . . be less inclined to tumultuous proceedings in seasons of scarcity, and . . . at all times be less influenced by inflammatory and seditious publications."[5] In order to persuade the lower and reassure the higher classes that despite inevitable poverty and want the existing order is the best possible, it was necessary to explain how the inconveniences occasioned by the principle of population were conducive to universal good. This Malthus undertook to do. Likewise, those natural theologians who accepted Malthus's principle were invariably able to accommodate it in their reconciling accounts of society and thereby to make it serve, as Malthus had intended, the interests of maintaining social order.

There is ample reason to suppose that Darwin was acquainted with the tradition of natural theological interpretation of the principle of population even before he read Malthus in 1838. Forty years after its first publication, Malthus's principle was still a topic of active debate and a problem on which natural theologians could exercise their ingenuity. It occupied several chapters in John Bird Sumner's *Treatise on the Records of the Creation*, with which Darwin was likely familiar.[6] It was alluded to briefly in William Kirby's Bridgewater treatise, which Darwin read in 1837–8.[7] And Paley devoted considerable attention to it in *Natural Theology*. Although Paley's view is not typical, being more optimistic than most, it is worth looking at simply because we know Darwin read it. It may, indeed, have been his first introduction to Malthus.

Paley discussed population under the heading "Goodness of the Deity." One purpose of the chapter was to prove that perceived evils are in fact good, that is, that they serve some useful purpose. The fundamental cause of one class of evil, the "evils of civil life," Paley said, is the principle of population, which he stated in Malthus's terms. The order of generation proceeds by a geometric progression, while the increase of subsistence assumes the form of an arithmetic series. The result of population's outstripping food supply is poverty, which imposes labor, servitude, and restraint. These things are inevitable. Even if there should be established a country where all people were "easy in circumstances," such a state of affairs would lead to an increase in marriages, and therefore children, which would soon cause scarcity and the necessity of toilsome labor. It does not follow, however, that the tendency to increase is an evil, for one consequence of it is that whenever the condition of a people is meliorated, either "the *mean* happiness will be increased, or a greater number partake of it; or, which is most likely to happen, that both effects will take place together." (Although this has a considerably more hopeful ring than Malthus's own statement of the case, it should be recalled that Malthus himself believed we could rationally expect a "gradual and progressive improvement in human society.")[8] Paley explained further that the increase of population is a necessary result of that part of human nature "which no one would wish to see altered," namely, the urge to marry and procreate. It is the result of "the provision which is made in the human, in common with all animal constitutions, for the perpetuity and multiplication of the species," which is obviously good.[9]

Elsewhere in the same chapter, in a slightly different context, Paley treated population chiefly as it relates to animal life. His problem was to explain away an apparent evil, the devouring of one animal by another – "the chief, if not the only instance, in the works of the Deity, of an economy, stamped by marks of design, in which the character of utility can be called in question." Paley found several reasons why this is not really an evil, of which the foremost is related to a "property of animal nature, viz., *superfecundity*." It is necessary that the prolific faculties of animals be curtailed, and animals' preying on each other serves this purpose. To be at all convincing, the argument must of course show why superfecundity itself is useful. Paley indicated two advantages of it. The first is that it tends to keep the world full. It may seem from man's limited perspective that there can be no good in the multiplication of annoying insects and other pests. But without it, he reasoned, "large portions of nature might be left void": "If the accounts of travellers may be depended upon, immense tracts of forests in North America would be nearly lost to sensitive existence, if it were not for *gnats* . . . Thus it is, that where we looked for solitude and deathlike silence, we meet with animation, activity, enjoyment; with a busy, a happy, and a peopled world." Plagues of destructive mice can, naturally, be accommodated in the same explanation; and Paley gloried in the consideration that what to humans appear to be blights are often hordes of animals "claiming their portion in the bounty of Nature."[10]

The second advantage of superfecundity is that it makes possible fluctuations in populations of different species as circumstances require. When the American forest regions are improved and "our gnats" disappear, other inhabitants can quickly take their place. It is under this second heading that Paley addressed the question of the utility of populousness among humans:

> It may be a part of the scheme of Providence, that the earth should be inhabited by a shifting, or perhaps a circulating population. In this economy, it is possible that there may be the following advantages: When old countries are become exceedingly corrupt, simpler modes of life, purer morals, and better institutions, may rise up in new ones; whilst fresh soils reward the cultivator with more plentiful returns. Thus the different portions of the globe come into use, in succession, as the residence of man; and, in his absence, entertain other guests, which, by their sudden multiplication, fill the chasm.

For man and beast, superfecundity is good. Therefore the devouring of one animal by another is good. "Though there may be the appearance of failure in some of the details of Nature's works, in her great purposes there never are." For those who might not have been convinced by his particular arguments, Paley added the general proposition that such evils as exist are necessary parts of a beneficent system of natural laws.[11]

The principle of population was for Paley by no means destructive of the harmonious view of nature. Nor was it for Malthus. Darwin may never have devoted much thought to the optimistic arguments of Paley and others concerning superfecundity. But when he read Malthus he found, side by side with the principle of population, an explanation of the good consequences that follow from it. In the first edition of his *Essay* Malthus devoted the two concluding chapters to natural theological reasoning on his principle, an attempt, as he put it, to give a view of the situation of man on earth that is consistent with our ideas of the "power, goodness, and foreknowledge of the Deity." The trials of this life are God's means of creating mind out of inert matter, of sublimating the dust of the earth into soul, he argued. If it were not for the wants of the body, there would be no stimulus to exertion and man would be sunk to the level of the brutes. The world would not have been peopled or the earth cultivated. The difficulties occasioned by the law of population contribute to that variety of impressions which is favorable to the growth of mind; they generate talents; they awaken dormant faculties. Thus, "there is no more evil in the world than what is absolutely necessary as one of the ingredients in the mighty process." The evil that exists is for the purpose of creating not despair, but activity.[12]

In subsequent editions these chapters on "the formation of mind" were dropped and Malthus's natural theology distributed into several other parts of the essay. The bulk of it came to rest in Book IV, Chapter I, where Malthus introduced moral restraint (postponement of marriage) as a means of avoiding the "positive checks" to population – misery and vice. The human activity which produces all the advantages and improvements of civilized life, he argued, is the result of the powerful and universal desire for food, and for clothing, houses, and so forth. But this desire alone is an insufficient stimulus to the sort of actions that the creator's purposes require:

> The desire of the means of subsistence would be comparatively confined in its effects, and would fail of producing that general activity so necessary to the improvement of the human faculties, were it not for the strong and universal effort of population, to increase with greater rapidity than its supplies. If these two tendencies were exactly balanced, I do not see what motive there would be sufficiently strong to overcome the acknowledged indolence of man, and make him proceed in the cultivation of the soil. The population of any large territory, however fertile, would be as likely to stop at five hundred, or five thousand, as at five millions, or fifty millions. Such a balance therefore, would clearly defeat one great purpose of creation.

If it be objected that the balance is shifted too far to the side of population, he continued, and is thus the cause of great misery, this is a question of degree; and we are not competent to judge of the exact strength of the tendency to increase that is necessary to achieve the ends of the creator with the smallest amount of incidental evil.[13]

It is evidently this discussion to which Darwin was referring when he wrote that population pressure in nature would "do that for form, which Malthus shows is the final effect . . . of this populousness on the energy of man." For Malthus, the potential increase of population beyond the means of subsistence produces that human activity which is required to accomplish the creator's aims. For Darwin, similarly, the force of population produces that constant adaptation of organisms to conditions which he believed to be characteristic of "this perfect world." Influenced by the Darwin of the *Origin of Species,* we have become accustomed to thinking of natural selection as necessarily implying a universe of chance and imperfection. But in fact there is in Darwin's notebook entries for September 28, 1838, no suggestion that he had ceased to believe in a "system of great harmony."[14] Instead we find a direct analogy drawn between Darwin's new theory and the arguments that Malthus used to incorporate the principle of population into that very system. From the fact that Darwin drew such an analogy it is a fair inference that he did not immediately see disharmony as the necessary concomitant of transmutation by means of the warring of species. Examination of his notebooks and some of his reading notes for the period of several weeks after he read Malthus supports and strengthens this conclusion.

In the pre-Malthus period of his speculations on transmutation, Darwin assumed that the creator established a hierarchy of laws, including those by which the adaptation of organisms to external conditions is maintained. Both this conception and Darwin's constant search for the final causes of everything reflect his belief that there is a plan of creation, that the laws of nature are the creator's way of achieving his desired ends. In this Darwin was in agreement with most of his contemporaries. During the spring and summer of 1838, he carried this conception of law and plan very far, denying free will and arguing that the human mind, like every other part of nature, functions in a lawful manner.[15] Thoughts, ideas, and feelings are due to "structure of brain." "Love of the deity effect of organization, oh you materialist!" he exclaimed.[16] Those who have seen in Darwin's "materialism," particularly in the statement just quoted and others like it, either atheism, agnosticism, or a departure from the idea of a plan of creation have simply misread him.[17] What he was saying is that man loves God because God established a system of laws that has as one of its purposes the production in man of "love of the deity." The following passage makes this abundantly clear:

savages . . . consider the thunder & lightening the direct will of the God[s] (& hence arises the *theological* age of science in every nation according to M. le Comte). Those savages who thus argue, make the same mistake, more apparent however to us, as does that philosopher who says the innate knowledge of creator has been implanted in us (individually or in race?) by a separate act of God, & not as a necessary integrant part of his most magnificent laws, which we profane in thinking not capable to produce every effect of every kind which surrounds us.[18]

Although Darwin's "materialism" would have been unacceptable to most of his contemporaries, his proposition that man's innate knowledge of a creator, along with every other effect of every kind which surrounds us, is a necessary result of God's laws nevertheless provides a striking illustration of his adherence to the common view that nature was preplanned by an omniscient creator.

Darwin did not abandon this way of thinking as soon as he conceived of natural selection. Implicit in his first statement of the theory is the question that characterizes his earlier speculations: What is the final cause of the law of population? The answer is that it serves to adapt organisms to a changing world. It is a part of the plan.[19] At about the same time, Darwin reiterated the idea of a system of "laws invoking laws & giving rise at last even to the perception of a final cause."[20] In the fourth transmutation notebook there is a long passage, written just over a month later, in which Darwin expressed unequivocally his belief that the existence of man is a part of the creator's plan. The immediate question under discussion is the "final cause" of separate sexes. One reason they exist, Darwin argued, is so that by the blending of slight differences the uniformity of species can be maintained. Without sexes all individuals would vary in different ways in accordance with the different slight changes in conditions that affect each, and there would then "be as many species, as individuals," which "is not the order in this perfect world." Such a situation would make impossible all social animals, including man. And since "man is *one* great object for which the world was brought into present state," sexes are a necessary part of the system.[21] The same ideas of the perfection of the world and the purposiveness of the creator's laws appear in Darwin's extensive commentary on the geologist John Macculloch's *Proofs and Illustrations of the Attributes of God*. Darwin there insisted that the creator works not "DIRECTLY," but through a "GREAT SYSTEM" of "grand & simple laws," and that particular aspects of the system demonstrate his wisdom. Who will dare to say that extinction is "an infringement on the wisdom . . . [of] Providence" when rocks and mountains are composed of the remains of dead and extinct forms, he asked.[22]

The question may be raised at this point how Darwin could continue to believe in a plan of creation for some two months after he adopted the theory of natural selection. In the theory of natural selection variations are "accidental," by which Darwin meant not that they have no determinate cause, but that they bear no relation to the organism's conditions of life; they are not adaptations. They are mere differences, the most appropriate of which have the best chance of survival. The idea of accident is not readily compatible with the idea of a detailed plan, such as Darwin still held in October and early November. When, then, did accidental variations become a part of the theory? The fourth transmutation notebook (October 1838–July 1839) indicates that this did not occur until some weeks after Darwin read Malthus and that in the interval he still held, almost unchanged, his old theory. On October 4 he wrote that "change of form is solely adaptation of whole of one race to some change of circumstance." There is no hint here of selection, "picking," or the wedging force of population. Rather, the reference to "whole of one race" seems to imply that all vary in the right way. A passage written two or three weeks later also sounds like the pre-Malthus Darwin. Conditions that are unfavorable for the parent may be favorable for its young because of "influence on parent affecting offspring," he said. Again, there is no suggestion of selection, but rather of an automatic organic response to change. The example given in support makes this interpretation still more probable. In dogs transported to Australia, the conditions of life and the habits of the parents produce mottled coloring in the puppies: "tendency in manner of life to be mottled & hereditary *tendency* determines the puppies to be so."[23]

During the first week in November, in the continuation of the passage already referred to, in which man is said to be a preplanned goal of the process of transmutation, Darwin restated, briefly but fairly completely, his pre-Malthus theory:

Without sexual crossing, there would be endless changes, & hence no feature would be deeply impressed on it, & hence there could not be *improvement,* & hence not in higher animals – it was absolutely necessary that Physical changes should act not on individuals, but on masses of individuals – so that the changes should be slow & bear relation to the whole changes of country, & not to the local changes – this could only be effected by sexes.[24]

Here Darwin is still saying that every change in conditions, however slight, calls forth corresponding changes in organisms. All the organic changes are appropriate for ("bear relation to") the external changes, and without sexes, organisms would, as Darwin said earlier, "become adapted to every minute change." But the minor alterations that "bear relation to" mere local changes are blended out to make possible adaptation to the "whole changes of country."[25]

The earliest passage I have seen in the transmutation notebooks in which variations are unambiguously stated to be mere differences, rather than adaptive changes, is from early March 1839: "my principle [is] the destruction of all the less hardy ones & the preservation of *accidental* hardy seedlings: . . . to sift out the weaker ones: there ought to be no weeding or encouragement, but a vigorous battle between strong & weak." By this time chance variation has clearly become a central part of Darwin's theory. On November 27, 1838, by contrast, the survival of chance variants was merely one possible explanation of change.[26] Darwin's notes on Macculloch's *Attributes of God* show signs of having been written during the period of transition in his thinking. While continuing to speak of the "GREAT SYSTEM" and the wisdom of providence, he asked whether he ought to use final causes as a guide to his speculations.[27] And in part of his notes he discussed the differential probability of survival of puppies in a single litter: "Suppose six puppies are born.[28] & in the Malthusian rush for life, only two of these live to breed. if circumstances determine that the long legged one shall rather oftener than any other one survive in ten thousand years the long legged race will get the upper hand, though continually dragged back to old type by intercrossing with ordinary race."[29] Early in December, probably as a result of reflecting on the role accident was assuming in his theory, Darwin reversed his previous position on man, saying that he was a "chance" production rather than "*one* great object" of the creator's plan.[30]

If it was not until toward the end of November that Darwin came to see variation as not automatically adaptive, then what exactly was the function of the wedging force of population in Darwin's theory during the previous several weeks? How did it serve to "sort out proper structure, & adapt it to changes"? If we return to the notebook entry for September 28, we find one suggestive expression. The wedges, Darwin said, "force every kind of *adapted* structure into the gaps in the œconomy of nature." The implication seems to be that for Darwin adaptation was still an absolute matter. Some forms are well adapted, and they are forced into the gaps. Others are not, and they perish. Since Darwin continued to assume that variation tends to be adaptive, wedging must principally have served to eliminate the unfortunate cases in which the generative system fails to produce the right change. The thrusting out of "weaker ones" would then have reference chiefly to these misfits and to the parent forms, which, as a result of the change in conditions, are no longer well adapted. Malthus, in other words, gave Darwin a greatly improved version of the traditional notion of selection in the sense of "nature's broom," a means of eliminating the not well adapted. But the wedging force of population did not at once take over entirely the function of producing adaptation. Rather it favored the survival of the well-adapted variations produced after changes in conditions.[51]

The following passage, from around October 10, suggests the same conclusion: "no structure will last without it is adaptation to *whole* life of animal, & not if it be solely to womb as in monster, or solely to childhood, or solely to manhood – it will decrease & be driven outwards in the grand crush of population. – "[52] This is the old assumption of adaptation by generation, but in a state of transition. Darwin still supposed that the generative system tends to produce adaptations, but they are not always adaptations for the whole life of the organism. Variations that are well adapted for only a portion of an individual's life perish in the Malthusian struggle.

One problem facing Darwin all through the first three transmutation notebooks was to know what causes variations to be adaptive. Until September 1838 his solutions were all internal organic mechanisms. In response to environmental changes, organisms alter in ways that adapt them and their offspring to the new conditions. They do so by means of their reproductive systems and the physiological laws that produce structural changes to correspond to changes of habit. Malthus's population pressure gave Darwin an external cause of adaptation, the wedging action that forces organic forms into gaps in the economy of nature.[53]

But for this truly to be the *cause* of adaptation, variations must be mere differences which the wedging force molds into new perfectly adapted forms. In the few weeks after reading Malthus, Darwin gradually adopted this view and abandoned his long-held belief that adaptations are born. The idea that variations have no adaptive relation to conditions appears to have been a consequence of, rather than a prerequisite for, Darwin's adoption of Malthusian views. If this interpretation is correct, that the transformation of Darwin's understanding of variations was produced gradually in the two months after he read Malthus, then his continued belief in a plan of creation during this period is readily intelligible. His new theory at this point was little more than the old but with population pressure grafted onto it to explain better the extermination of the unfit. Lacking as yet the essential idea of accidental variations, it was still compatible with belief in a preplanned system.[34]

Darwin in his autobiography recalled that when he wrote the *Origin of Species* he believed in "a First Cause having an intelligent mind in some degree analogous to that of man" and that this cause, not "blind chance or necessity," was responsible for this "immense and wonderful universe." He deserved, he said, "to be called a Theist."[35] I know of no evidence that before 1859 Darwin ceased believing in a harmony of a very general sort – that the result of the creator's laws is on the whole good. Indeed, the similar conclusions to his "Sketch" and "Essay" of the 1840s and to the *Origin* closely resemble the theodicy of Malthus. "From death, famine, and the struggle for existence," he said, "we see that the most exalted end which we are capable of conceiving, namely, the creation of the higher animals, has directly proceeded."[36] John Greene's characterization of Darwin as an evolutionary deist is clearly more appropriate, for both the pre- and the post-Malthus periods, than the designations "agnostic" and "atheist" that have sometimes been employed.[37] Yet Darwin's views had indeed changed substantially. Within a month or two of reading Malthus, Darwin concluded that the natural selection of accidental variations could not be reconciled with the idea of a plan of creation.[38] From then on he assumed that the creator imposed "only general laws."[39] His view, as he reported it later to Asa Gray, was that there is no exact plan; that everything is the result of "designed laws," but that the details are left to the working out of "chance."[40] His new theory was consistent with a belief in preordained results of a general nature, such as the production of adaptation or of the higher animals, but not, at least very readily, with the proposition that any particular organ or organism was preordained. This perhaps should be taken as marking the end of Darwin's adherence to the traditional conception of the harmony of nature. At the least it is a substantial departure from the conception of a divine plan.[41] After December 1838 references to the "GREAT SYSTEM" disappear from Darwin's writings.

The structure of Darwin's theory, 1838–1844

Darwin's rejection of the idea of a world preplanned in all its particulars represents a significant divergence from the views of most of his fellow naturalists. Henceforth accident would play a role in his thinking that few of his contemporaries were prepared to admit into theirs. But for a long time Darwin continued to hold in common with them some of the main assumptions of the harmonious conception of nature. Most importantly, in the earliest expositions of his theory, the "Sketch of 1842" and the much longer "Essay of 1844," he still believed in perfect adaptation, rather than the relative adaptation we find in the *Origin of Species*.[42]

Several strands of evidence lead to this conclusion. First, there is the fact that Darwin on numerous occasions in the "Essay" indicated that he considered "perfect adaptation" to be the norm in nature. Unless there is some change in external conditions to disrupt adaptation, organisms are perfectly suited for the situations in which they live. Geological changes, he said, result in organisms' ceasing to be "so perfectly adapted . . . as they originally were." When an island is in process of formation and some plants or animals by chance arrive there, "it is impossible that the first few transported organisms could be perfectly adapted" to all the stations on the new island; "and it will be a chance if those successively transported will be so adapted. The greater number would probably come from the lowlands of the nearest country; and not even all these would be perfectly adapted to the new islet whilst it continued low and exposed to coast influences." While an archipelago is forming, he said, its stations will not at first be occupied "by perfectly adapted species."[43] In all these instances perfect adaptation is treated as what is to be expected *except* when conditions are altered.

There are, I should point out, other places in the "Essay" in which Darwin might seem to deny perfect adaptation. In reading these, however, it is necessary to keep in mind the distinction between the teleologists' conception of perfect adaptation and the idea of limited perfection that Darwin, Owen, and others espoused. To the teleologists a perfectly adapted form was the best possible form for the conditions under which it lived. Those who rejected teleological explanation, on the other hand, held that the perfection of organisms was limited by certain general laws, such as the law of adherence to a common type, or, in Darwin's case, the laws of heredity.[44] The idea of limited perfection is nicely expressed in some of Darwin's early post-Malthus notes: "I look at every adaptation, as the surviving one of ten thousand trials – each step being perfect – or nearly so (except in isl[d]) although having hereditary superfluities. Man could exist without mammae – to the then existing conditions."[45] Perfection here is limited by the "hereditary superfluities." In the pre-Malthus notebooks Darwin constantly argued against the teleologists' explanations, and he continued to do so in the "Essay of 1844." It is in this context alone that Darwin denied perfect adaptation, and it is the teleological concept of perfection that he denied. For instance, in discussing geographical distribution, Darwin was concerned to show that it could not be explained teleologically, that is, solely in terms of the adaptation of organisms to external conditions. To prove the teleological explanation inadequate, he cited the case of introduced forms that compete successfully against the native inhabitants:

> Although every species is admirably adapted (but not necessarily better adapted than every other species, as we have seen in the great increase of introduced species) to the country and station it frequents; yet it has been shown that the entire difference between the species in distant countries cannot possibly be explained by the difference of the physical conditions of these countries.[46]

This is much like the argument Agassiz would later use against the teleological interpretation of the facts of geographical distribution. And it is precisely the same argument Darwin employed as early as the first notebook: "a Race of domestic animals made from influences in one country is permanent in another – Good argument for species not being so closely adapted."[47] It is a telling blow against the teleologists' version of perfection, but it leaves untouched Darwin's own, that within the limits imposed by the laws of heredity and the accidents of transportal, the native inhabitants are perfectly adapted. In Darwin's theory of 1844 the natives cannot be further improved because, being perfectly adapted, they do not vary. Only if some change in conditions causes them to be no longer perfectly adapted will natural selection begin to work. How far Darwin still was in 1844 from the idea that species are only relatively well adapted, in comparison with their competitors, can be seen by contrasting his conclusions on introduced forms in the "Essay" with those he drew in the *Origin of Species*. In 1844 he said, "I mean by not being perfectly adapted, only that some few other organisms can generally be found better adapted to the country than some of the aborigines." In the *Origin* he added, significantly: "We may safely conclude that the natives might have been modified with advantage, so as to have better resisted such intruders."[48] In 1859 Darwin interpreted the phenomenon to mean that the natives were not perfectly adapted in any meaningful sense. There is always room for improvement. In 1844 it meant only that the teleologists' principle of adaptation was untenable, thus leaving the door open for Darwin's rival explanation in terms of transmutation and transportal.

A second strand of evidence comes from Darwin's description of the results of natural selection. "My theory," he wrote, shortly after reading Malthus, "makes all organic beings perfectly adapted to all situations where in accordance to certain laws they can live."[49] The same idea occurs in Darwin's comparisons of man's selection with nature's. This is a topic that first appeared in the E notebook in December 1843. "If nature . . . had the picking" she could make a web-footed dog "far more easily than man." Domestic races, Darwin said, "are made by precisely same means as species – but latter far more perfectly & infinitely slower." No domestic animal, however, "is perfectly adapted," he added, implying that natural species are.[50] In the "Essay of 1844" Darwin contrasted the imperfection of man's techniques with the operation of the "natural law of selection." Man, he said, selects chiefly by eye and cannot see whether the internal structure is well suited to the external. Nor can he detect shades of constitutional differences. He has bad judgment, and he is capricious: He and his successors do not select for the same end for hundreds of generations. Man selects forms that are useful to him rather than those that are best adapted to the conditions under which the animal lives. And he does not keep those conditions uniform. Frequently he is unwilling to destroy an animal that is not up to his standard. And he often begins selecting with a sport that departs considerably from the parent form.

Nature, on the other hand, selects varieties that differ only slightly from the parent forms. Conditions are constant for long periods and change only slowly. Crossing rarely occurs. Nature never fails to destroy forms that are not well adapted and never selects forms unless they are better adapted than their parents. Nature's selection is not capricious, but goes on steadily for thousands of years adapting the form to the same conditions. The selecting power works on internal structures as well as external, testing the whole organism during its whole life, and if it proves to be less well adapted than its congeners it is inevitably destroyed. "Every part of its structure is thus scrutinised and proved good towards the place in nature which it occupies." It follows that nature's productions will be far more perfect than man's. In domesticated animals, selection for the same ends, steadily, for a number of generations, under suitable conditions, and without crossing, produces forms that are "true," or subject to little variation. Nature's forms must be incomparably "truer," for they have been made by rigid and steady selection, without crossing, continued for thousands of years. The result is forms "excellently trained and perfectly adapted" to conditions. Such forms seldom vary, which Darwin believed to be one sure mark of a species.[51]

Darwin's conception of the process by which these perfectly adapted forms are produced rests on the idea that "accidental" variations are differentially adaptive. When geological changes alter the conditions under which a species lives, or when individuals of a species are transported to a new region, the reproductive system is affected, and the species begins to vary. Almost every part of the body, Darwin said, would tend to vary in slight degrees and in no determinate way.[52] The variations, that is, are not automatically adaptive. Some variant individuals will be better adapted, others "less well adapted," to the new conditions, and their chances of survival will be different. The process of selection, once begun, continues until (and only until) it has turned out forms as perfect for the conditions as their hereditary structure allows. When such forms have bred for many generations under the new conditions, the alterations they have undergone become fixed in their constitutions and they breed true.[53] It is important to note that the idea of differential adaptiveness of variations is not equivalent to the idea that species are only relatively well adapted in comparison with their competitors.[54] In Darwin's theory of 1844 it is assumed that some variant individuals that occur under changed conditions will be better adapted than others and than the parent form, which as a result of the change ceases to be perfect. But species, the stable, true-breeding end products of selection, are perfectly, not merely relatively well, adapted. Natural selection gave Darwin a new explanation of how adaptation is produced, but it did not, until after 1844, lead him to a new view of the final product.[55]

The evidence discussed so far suggests strongly that Darwin as late as 1844 still believed in perfect adaptation, but taken by itself it is not conclusive. Even after 1859, when Darwin definitely considered adaptation to be a relative matter, he occasionally slipped into the language of perfect adaptation, and at times he seems consciously to have employed it for polemical purposes.[56] In 1844 too was he perhaps simply continuing to use the term after he had already abandoned the concept? Careful examination of Darwin's theory as presented in the "Essay" reveals that this is not the case. It shows that Darwin still believed in perfect adaptation in a meaningful sense and that this belief determined the structure of the theory at this stage in Darwin's thinking. In the "Essay," natural selection works only intermittently. Its action is constrained and limited by the only occasional availability of variations. And variation is limited because it is produced only when organisms cease to be perfectly adapted. In 1844, as in the earliest period of his speculations, Darwin assumed that the generation of variations is regulated by the perfection of adaptation. As a result of this assumption, in the "Essay" the theory of natural selection is constructed on, and limited by, an essentially natural theological foundation.

Between 1837 and 1844, Darwin's theory of the generation of variations did undergo one major modification, as we have seen. In the first notebooks, Darwin suggested that external changes would destroy adaptation if they were not accompanied by corresponding organic changes. The balance was preserved, he thought, because external changes, by affecting the reproductive systems of organisms, cause them to produce offspring that are adapted to the new conditions. After he adopted the theory of natural selection, Darwin concluded that the generative process produces not new adaptations, but merely indifferent variations, on which natural selection works. The effect of this new view of variation was to remove one major function from the domain of the generative system. No longer would it be responsible for the production of adaptation. But in most other respects, Darwin's theory of generation remained the same. Darwin continued to believe that offspring unlike their parents are produced only when the reproductive system has been influenced by external changes. It is here that the concept of perfect adaptation enters the theory of natural selection and assumes the role of regulator of Darwin's mechanism. Until some time after 1844 the perfection of adaptation governed the production of variations in Darwin's theory, and in consequence it governed the entire process of organic change.

In the "Essay of 1844," as in the earlier "Sketch," a change in the conditions of life is Darwin's explanation of variation under domestication and in nature. Domestic variation is taken up first. Darwin explained that under certain conditions organisms are altered during their individual lives and that these alterations may be passed on to future generations. Habits of body and mind, structural changes resulting from use and disuse of organs, the effects of food and climate and occupational diseases, and the tendency to contract some diseases may all be inherited. But these variations due to the direct effect of conditions on individuals are, he said, "extremely rare compared with those which are congenital or which appear soon after birth." These latter are "infinitely numerous," and it is chiefly on them that the breeder works to make new races. The cause of these "congenital" variations, he said, is "the accumulated effects of a change of all or some of the natural conditions of life of the species, often associated with an excess of food."[57] The changes act not directly on the organism, but on its reproductive system. This indirect effect causes the reproductive powers to fail to produce young exactly like their parents. The organization of the embryo becomes "in a slight degree plastic," and variations result.[58] Given Darwin's well-known familiarity with the writings and opinions of plant and animal breeders, it would be easy to assume that this was the source of his explanation of variation under domestication. In fact, however, it was not. His search of the literature turned up only two explanations, both offered by Andrew Knight and both unsatisfactory to Darwin. Darwin rejected outright Knight's suggestion that a more genial climate might induce variation, noting that plants from warmer regions vary in England. And although he agreed that an increase of food might often be a factor, he said it was insufficient to account for most variation.[59] Darwin's insistence on the importance of external change was derived not from the breeders, but from his own view of variation in nature. This view is the embodiment, in his theory of generation, of the assumption that we find in the opening pages of the first notebook – that the purpose of variation is accommodation to a "changing world."

Darwin's reason for beginning with his theory of domestic variation was that only in this way did he think he could make a plausible argument for the existence of variation among animals and plants in a state of nature. He believed that there is very little variation in the wild and that consequently he could not simply go to the writings of naturalists, as he did to those of the breeders, to prove that there is sufficient material for selection to work on. But by arguing that the recognized variability of domestic plants and animals is caused by a change in conditions, he could conclude that there must be variation in nature as well, for geology teaches that there is a constant round of changes in climate and other environmental conditions:

> Domestication seems to resolve itself into a change from the natural conditions of the species . . . ; if this be so, organisms in a state of nature must *occasionally*, in the course of ages, be exposed to analogous influences; for geology clearly shows that many places must, in the course of time, become exposed to the widest range of climatic and other influences.[60]

The slow changes from geological causes probably act on the reproductive organs during several generations before the accumulated effects cause the species to vary. Sometimes, however, a Lyellian catastrophe ("as when an isthmus at last separates") or the dissemination of seeds into a new region will suddenly bring an organism into contact with new conditions and thus produce a tendency to vary. "The reproductive system would be affected, as under domesticity, and the structure of the offspring rendered in some degree plastic"; or, as he said in 1842, "the mould in which they are cast" would vary slightly.[61] The hereditary type or "mould," fixed and stabilized by the long-continued process of generation, would cease, under altered conditions, to govern rigidly the form of new individuals.

In the "Essay," as in his pre-Malthus theory, Darwin supposed that even in the absence of significant external change, individuals of the same species would often not be exactly similar: "the recognition . . . of one animal by another of its kind seems to imply some difference," he said. Before reading Malthus, Darwin considered these slight differences to be adaptive changes caused by local fluctuations in external conditions, and he thought they were blended out to allow for adaptation to more general and long-continued changes. In the "Essay" too they are said to be caused by slight changes, but they are now mere differences, rather than adaptive variations.[62] They may be blended out, or they may be selected; but though selectable, they are not the variations on which Darwin based his theory in 1844. This is in sharp contrast to the *Origin*, in which natural selection does indeed rest squarely on "individual differences." By 1859 Darwin believed that even in the absence of any great change in conditions, individuals may differ in such important respects as the arrangement of the central nervous system. "Individual differences" might include variations in any part of an animal's organization.[63] In 1844 his opinion was quite different. He then thought that without significant external change, "the variation, such as it is, chiefly affects . . . the size, colour, and the external and less important parts"[64]; that is, those parts which, according to his theory of generation and his law of hereditary fixing, are most affected by external change and are the least firmly fixed, the most easily varied. The reason Darwin did not base his theory on such variations is obvious. The great changes that have marked the history of most groups of organisms could not have been the result of variations only in the "external and less important parts." Darwin required variations that could affect "every part of the body," including those most centrally important and most firmly fixed.

Accordingly, in the "Essay" two kinds of variation are discussed. There is the "small and admitted amount of variation which has been observed in some organisms in a state of nature," and there is the "hypothetical variation consequent on changes of condition."[65] No "just distinction" can be drawn between them, Darwin said, because both are caused by external change – one by slight local changes, the other by more permanent and perhaps greater changes – and because the fixity of characters is a matter of degree. But though a matter of degree, the difference between the two kinds of variation was, in Darwin's view, significant. One paragraph in the "Essay" suggests the greater importance which he attached to the second. Let us suppose, he said,

> that an organism by some chance (which might hardly be repeated in 1000 years) arrives at a modern volcanic island in process of formation and not fully stocked with the most appropriate organisms; the new organism might readily gain a footing, although the external conditions were considerably different from its native ones. The effect of this we might expect would influence in some small degree the size, colour, nature of covering &c., and from inexplicable influences even special parts and organs of the body.

Though introduced here as part of an example of the effects of a definite change in conditions, these are the same slight variations in external parts that are often found among individuals.[66] "But," Darwin continued, "we might further (and [this] is far more important) expect that the reproductive system would be affected, as under domesticity, and the structure of the offspring rendered in some degree plastic. Hence almost every part of the body would tend to vary from the typical form in slight degrees, and in no determinate way."[67] As under domesticity, the "whole frame" might vary, rather than just the less important parts.

Darwin's theory in 1844 required that occasionally organisms be subjected to distinct changes in conditions.[68] The changes need not be catastrophic or of great magnitude, though they might be either; but they must be definite. They must be more than minor environmental fluctuations, for only a significant change is able to unsettle the fixed hereditary constitution of a species. This belief in the necessity of external change to produce variation, which Darwin carried over virtually unaltered from his pre-Malthus period, had two important consequences for the theory of natural selection. One is that it sharply limited the availability of materials on which natural selection might work. Geological conditions remain the same for long periods of time.[69] And geological change is slow, so that only after a long time will its accumulated effects

81

cause a species to vary. Therefore, there can be very little variation among organisms in a state of nature. This Darwin stated again and again in the two essays of the 1840s. "Wild animals vary exceedingly little"; "the proverbial expression that no two animals or plants are born absolutely alike, is much truer when applied to those under domestication, than to those in a state of nature"; "most organic beings in a state of nature vary exceedingly little"; "the amount of variation [is] exceedingly small in most organic beings in a state of nature, and probably quite wanting (as far as our senses serve) in the majority of cases."[70] This conclusion followed directly from Darwin's belief that in the absence of external change organisms are perfectly adapted and so do not vary.

The second consequence, closely related to the first, is that transmutation can occur only in those rare situations in which there is a change in conditions. This is not explicitly stated in the "Essay of 1844." But in both the "Essay" and the "Sketch of 1842," whenever he discussed variation or gave an illustration of the working of natural selection, Darwin stipulated that there has been a change in conditions.[71] At the end of the "Sketch" he went so far as to leave himself a note on the need to include a discussion from Lyell, or a reference to him, "to show that external conditions do vary" – a fact that virtually every geologist and naturalist in Darwin's day fully admitted.[72] In the section "Natural means of selection" in Chapter II of the "Essay," Darwin indicated most clearly his view of the difference between the situation in which conditions are changing and that in which they remain the same. The Malthusian principle operates continually to produce struggle in nature. But with "the external conditions remaining the same," the result of the struggle is not change, but stability: By struggling with individuals of the same or different species, each individual of each species merely "holds its place." The struggle presumably serves to keep the species up to its standard of perfection by weeding out misfits. However, when "the external conditions of a country change," the original inhabitants cease to be so perfectly adapted as formerly; the change affects their reproductive systems, making the organization of their offspring plastic; and in the struggle for subsistence, chance favors those variant individuals that are better adapted to the new conditions.[73]

In the "Essay of 1844" natural selection is not, as it would become in the *Origin of Species,* an ongoing process, working constantly at the improvement of organisms. Like the mechanisms of Darwin's pre-Malthus theory, it operates only as an organic response to changed conditions. The reason for this is that Darwin continued to assume that adaptation is perfect, an assumption that is deeply embedded in his theory of variation. In the fourth transmutation notebook, the correlation between less-than-perfect adaptation and variability is stated explicitly: "No domesticated animal is perfectly adapted to external conditions. – (hence great variation in each birth)."[74] Perfectly adapted organisms have no need to change, and so they do not vary. Another passage from the fourth notebook, written in March 1839, says the same thing: "In the place where any species is most common, we need not look for change, because its numbers show it is perfectly adapted."[75] Seven years later Darwin's opinion remained unaltered. A note dated July 1846 reads: "When new form introduced into island, it must not be perfectly adapted, else it will not vary."[76]

Darwin's persistent belief in perfect adaptation dictated that the structure of his theory of transmutation in 1844 would be essentially the same as in the pre-Malthus notebooks. In the period of his early speculations, the purpose of transmutation was to preserve the harmonious adaptation of organisms to their environments. When external change threatened this adaptation, a natural mechanism was automatically called into play to maintain or restore it. The same pattern of change characterizes the "Essay of 1844." Before Malthus, Darwin's mechanisms were the generation of new adaptations and the effect of changed habits. In the "Essay of 1844," changes in conditions bring variation and natural selection into play. In both cases, the result is the same – the production of new, perfectly adapted forms.

1844 versus 1859

When he opened his first transmutation notebook Darwin believed that species change in order to adapt to changing conditions. The central problem he grappled with throughout the first three notebooks was to discover the means by which the organic world adjusted to change so that the harmony of nature was not disturbed. Malthus provided Darwin with a new solution to his old problem, and at the moment of its formulation the new solution occupied a preexisting slot in Darwin's conception of a changing but always harmonious nature – the slot reserved for the process by which perfect adaptation is maintained. The new theory differed from Darwin's earlier speculations and was superior to them in that it explained how change occurs, whereas the theory of generation simply asserted the fact, and that it could account for coadaptations, which the habit-structure theory was unable to do.[77] But from 1838 until after 1844, natural selection bore the marks of Darwin's early commitment to the harmonious conception of nature. This is so despite the fact that in these same years Darwin's conception of nature changed, from a preplanned "system of great harmony" to the effects of a few "designed laws" which have only by chance produced precisely those beings and organic relations that actually exist. As Darwin's view of nature was transformed, the idea of adjustment to change in order to preserve harmony ceased to be of great interest to him. In the early notebooks, it was the primary reason for, the final cause of, the process of transmutation, and it occupied a central position in his thinking. By 1844 Darwin was much more concerned with the process itself, and with its implications for the study of natural history. The idea of a self-adjusting harmonious system is nowhere mentioned in the "Essay of 1844." But as long as Darwin continued to believe in perfect adaptation, the theory of natural selection closely resembled his pre-Malthus theory and was by the same token very different from the theory we find in the *Origin of Species*.

To appreciate the changes that occurred in Darwin's thinking after 1844 it is helpful to compare directly his views on several key issues in the *Origin* and in the "Essay." The most important difference is in his conception of adaptation. Whereas in 1844 he believed adaptation to be perfect, by 1859 he supposed it to be relative: "Natural selection tends only to make each organic being as perfect as, or slightly more perfect than, the other inhabitants of the same country with which it has to struggle for existence."[78] Several other propositions characteristic of the "Essay" are derived from Darwin's belief in perfect adaptation. In 1844 he thought that a change in external conditions was necessary to produce variation and transmutation. By 1859 he assumed that no change was required.[79] In the "Sketch," and by implication in the "Essay," Darwin imagined that perfectly adapted organisms are "cast" in a stable "mould." The mold becomes "plastic" only when changes in conditions have caused the organism to be less than perfectly adapted.[80] In the *Origin* he stated, with his former self obviously in mind, that "no one supposes that all the individuals of the same species are cast in the very same mould."[81] In 1844 he thought there was "exceedingly little" variation in nature, while in the *Origin* he implied that variation, even in important parts, is common.[82]

One consequence of the cluster of beliefs about adaptation and variation that Darwin held in 1844 is that in the "Essay" the process of transmutation is intermittent.[83] It operates only in response to changes in external conditions, caused either by geological changes or the migrations of organisms. At the beginning of his evolutionary speculations Darwin believed, with his contemporaries, that variations are for the purpose of effecting accommodation to change; and from this belief he assumed – as Lyell and Cuvier would have agreed – that variations are caused by the changes to which accommodation is required. In 1844 he still believed that perfectly adapted forms, having no need to accommodate, do not vary. Unless there is a change in conditions to produce variation, individuals of a species differ only in unimportant respects. In the *Origin* the action of Darwin's mechanism for change is no longer constrained by the idea of perfect adaptation. Natural selection is potentially in continuous operation. Since species are not perfectly adapted, there is always room for improvement. At any time, with or without external change, variations may occur, and selection may work on them to produce new, better adapted forms. In 1859 Darwin could say that natural selection is "daily and hourly scrutinising, throughout the world, every variation"; that it is "silently and insensibly working, whenever and wherever opportunity offers, at the improvement of each organic being in relation to its organic and inorganic conditions of life."[84] This is the sweeping vision of the evolutionary process that became possible for Darwin after he abandoned perfect adaptation. Contrast it with the cramped conception of 1844: If there are variation and struggle in nature, then "new races of beings will – perhaps only rarely, and only in some few districts – be formed."[85] The contrast is not merely between unimportant differences in expression. It reflects a fundamental shift in Darwin's conception of evolutionary change. In 1844 the working of natural selection was governed by natural theological assumptions about perfection, adaptation, and variation. By 1859 these assumptions were no longer an important part of Darwin's view of nature.

In 1844 the theory of natural selection still had a long course of development to undergo before it acquired the form it has in the *Origin of Species*. The path this development would take was not one Darwin could see ahead of him nor its end one that he could consciously attempt to reach. The shape of the theory as it is presented in the *Origin* is the result of Darwin's gradually adopting a different perspective on adaptation and variation, a perspective of which he did not see the possibility until he had arrived at it. In the interval, he was not trying to modify natural selection or to achieve a new point of view on the pattern of organic change. He was fairly well satisfied with his theory, and his main concern was simply to construct the best argument for it that he could. For this purpose he undertook to collect evidence for his propositions on variation, heredity, crossing, and so forth. But the principal task he set himself, one that he early recognized as essential to the success of his enterprise, was to bring his theory into harmony with the best-established facts of natural history: to show that it could explain all of the most important phenomena studied by his contemporaries. His speculative activity, in consequence, was largely devoted to questions of geographical distribution, embryology, morphology, paleontology, and classification. Out of his work on these subjects, as an unexpected by-product, came the transformation of the theory of natural selection.

4

Part II of Darwin's work on species

At an early period in his speculations, months before he conceived of natural selection, Darwin planned to write a book on the subject of transmutation. One component of the projected work was to be Darwin's own theory; a second and more important part was to be devoted to the proof that transmutation has occurred and the application of the idea of descent to the explanation of the leading facts of natural history. In the spring of 1838 he wrote that there was scarcely any novelty in his theory and that "the whole object of the book is its proof." Earlier he had indicated more fully which subjects, because they were best explicable by transmutation, provided the best proof of it: "study gradation, study unity of type, study geographical distribution, study relation of fossil with recent. The fabric falls!" A few weeks before reading Malthus, Darwin claimed, in what seems to be a sketch for an introduction or preface, that the whole value of his book would lie in the second part, the proof and application of the doctrine of descent: "Seeing what Von Buch (Humboldt) G. St. Hilaire, & Lamarck have written I pretend to no originality of idea – (though I arrived at them quite independently & have used them since) the line of proof & reducing facts to law only merit if merit there be in following work."[1]

Darwin's pre-Malthus notions about the organization of his "work" are reflected in the "Sketch of 1842" and the "Essay of 1844." Both are divided into two parts, the first of which may be generally described as the presentation of the theory of natural selection, while the second is Darwin's mustering of the evidence for the doctrine of descent. Part I comprises three chapters or sections, one each on variation under domestication and the principle of selection, variation in nature and natural selection, and "difficulties," such as the gradual production of very complex organs and the variation and selection of instincts and other mental attributes.[2] In Part II, which in the "Essay" is headed "On the evidence favourable and opposed to the view that species are naturally formed races, descended from common stocks," chapters or sections are devoted to paleontology and the fossil record, geographical distribution, classification, morphology, embryology, and abortive or rudimentary organs. There is no explicit division into parts in the *Origin of Species,* but the same basic arrangement is preserved, Chapters I–VIII corresponding to Part I and Chapters IX–XIII (in the first edition) to Part II.

Part II occupies a relatively small part of Darwin's published writings – in the first edition of the *Origin,* only about one-third of the whole. But this was to some degree an accident of the circumstances of composition. In both the "Sketch" and the "Essay" Part II is considerably longer than Part I (106 printed pages to 76 in the "Essay"). If Wallace's letter of June 1858 had not interrupted Darwin in the midst of writing his big book, the second half would probably have been as large as the first, for Darwin intended to devote whole chapters to classification, morphology, embryology, and abortive organs, which in the *Origin* are lumped together in Chapter XIII. The best indication of the extent to which Part II was slighted in the *Origin* is the material on Part II subjects preserved in the "Black Box" at Cambridge University Library (now DAR 205; see "Note on manuscript citations" at front of book). The "Black Box" contains, among other things, the remains of the filing system that Darwin devised for his "huge pile of notes" on transmutation.[3] Most of the notes were used in the composition of Darwin's various books, including *Natural Selection,* and are now in many cases collected in bound volumes in Cambridge University Library. But the notes relating to the parts of Darwin's big book that were never written, except as brief chapters in the *Origin,* are still much as Darwin left them. The most important of these are two envelopes of notes on geographical distribution, one on paleontology and extinction, one on abortive organs, one on embryology, and one on classification (this last, for reasons to be explained in Chapter 7, is labeled "divergence").[4] Like the best-known Darwin notes, the transmutation notebooks, the notes in these six envelopes are a mixture of speculations, sketches of arguments to be used in the big book, comments on reading, and notations of useful facts and quotable statements from the works of other naturalists. Most are from the period 1838–59, although some are from the 1860s and 1870s. Darwin was to have used these notes in the preparation of the latter chapters of *Natural Selection* as well as in the "third work" of his projected multivolume series on transmutation. The first of these works, and the only one to appear, was *The Variation of Animals and Plants under Domestication,* in two volumes. The second work was to be on variation in nature, the struggle for existence, natural selection, and difficulties on the theory. The third was to be the application of the theory to the explanation of geological succession, geographical distribution, morphology, classification, and so forth.[5] This third work would probably have been a very large one, for the six envelopes contain no small quantity of notes: As a rough guess I would say that, even excluding the post-1859 material, they are approximately equal in amount to the B, C, D, E, M, and N notebooks combined.

Part II predates natural selection in Darwin's thought, and it continued to be a major part of Darwin's projected work. Moreover, from 1838 until at least the mid-1860s, when the idea of descent gained fairly wide acceptance, Part II was for Darwin in a fundamental sense the more important part of his work. Darwin was fond of the theory of natural selection, but his greatest concern was to establish the doctrine of descent. Two letters from the post-*Origin* period are testimony to the relative value he placed on natural selection and the general idea of evolution. In one from May of 1863, published in the *Athenaeum,* Darwin defended natural selection as the best theory available, but he said that whether one accepted it, Lamarck's theory, the theory of the *Vestiges,* or some other "signifies extremely little in comparison with the admission that species have descended from other species and have not been created immutable; for he who admits this as a great truth has a wide field opened to him for further inquiry."[6] To Asa Gray he said at about the same time, *"Change of species by descent . . .* seems to me the turning point. Personally, of course, I care much about Natural Selection; but that seems to me utterly unimportant, compared to the question of Creation *or* Modification."[7] For this question the Part II chapters are crucial, for it is in them that are found, in Darwin's words, "the arguments which alone have much weight" in favor of descent.[8] It is because the general question of descent was primary for Darwin that in the

"Sketch" he identified the second as "the far more important" part of the whole. And it was for the same reason that he could write, shortly after receiving Wallace's letter, that although all his "originality" would be "smashed," the value of his book would not be decreased, "as all the labour consists in the application of the theory."[9]

Darwin's insistence on the importance of Part II might be supposed an inducement to historians to show more interest in it than they have. But there are other, more compelling reasons for attending to it. Part II is Darwin's evolutionary transformation of biology, his synthesis of the results of mid-nineteenth-century natural history. Given the fact that Darwin's synthesis, apart from the theory of natural selection, was widely adopted by biologists after 1859, it is clear that those results are themselves a not insignificant component of late-nineteenth-century evolutionism. In his autobiography, Darwin said that "innumerable well-observed facts were stored in the minds of naturalists, ready to take their proper places as soon as any theory which would receive them was sufficiently explained."[10] But those "facts" were not as unconnected in the minds of his contemporaries as Darwin's remark suggests. They were already stated in the form of generalizations most of which could be translated with relative ease into evolutionary terms. It was in the second part of his work that Darwin translated them, and it is therefore by studying Part II that we see most clearly the relationship of Darwin's synthesis to the biological thought of his day. It is also in Darwin's work on Part II that we see most clearly the social character of Darwin's scientific activity. Darwin's theory did not develop in the isolation of his study. It took shape in response to the concurrent development of theoretical natural history in the hands of his professional colleagues, whose ideas, interests, and attitudes reached him through conversation, correspondence, and his voluminous reading.

Darwin and his contemporaries: the method of Part II

Darwin's use of the ideas, observations, and expertise of some of his contemporaries has long been noted and commented upon. Apart from Lyell, Hooker, and Huxley, those whose assistance to Darwin has attracted the most attention are the plant and animal breeders. Darwin, perhaps misleadingly, gave them a very prominent place in his account of his work during the period of the transmutation notebooks, and so his relations with them have been the subject of considerable discussion.[11] Information derived from breeders was undoubtedly important for Darwin's work, particularly on the laws of generation and heredity, although to put their contribution in perspective it should be noted that the large majority of individuals and works cited in *Natural Selection,* which is almost entirely Part I, the presentation of Darwin's theory, are professional biologists and their writings. For Part II, professional biologists were by an even larger margin Darwin's major source of facts and ideas.[12]

Sandra Herbert has recently insisted on the importance of the services rendered Darwin after the *Beagle*'s return by the professional biologists to whom he entrusted his collections. She observes that a pivotal issue for Darwin as he contemplated accepting transmutation was whether among the South American ostriches and the Galapagos rodents, mockingbirds, and finches there were in each group two or more separate species, or whether all the finches, for instance, were merely varieties. The decisions by John Gould, for the birds, and George R. Waterhouse, for the animals, in favor of species, along with the several similarities between South American fossil and recent forms that Owen established, convinced Darwin in about March of 1837 that species as well as varieties were formed by natural means.[13] The very important and more general suggestion that Herbert makes is that Darwin's scientific orientation was from an early period toward the professional scientific community, and that Darwin must be considered "as he considered himself, that is, within the context of contemporary science." The way Darwin organized his speculations, his pattern of publication, and his intellectual development are all explicable in part, she urges, by "Darwin's professionalism as represented by his commitment to science as a vocation and his dependence on the activities and judgments of his coworkers."[14] This I think must be carried still further, to cover the development of Darwin's theory in the 1840s and 1850s, and to include not only the activities and judgments of his coworkers but also the principal generalizations of mid-nineteenth-century professional biologists.

Most of the working relationships Darwin established with professionals in the months after the *Beagle*'s return lasted well past the period of the transmutation notebooks and the *Journal of Researches* (1839). For instance, while Owen was working on the *Beagle* fossils, he and Darwin met frequently, either in Owen's quarters at the College of Surgeons or at Darwin's house. In the 1840s Darwin continued to visit Owen in London, and Owen on occasion traveled to Down. A few letters passed between them. In the transmutation notebooks, the Black Box notes, Darwin's letters, and Owen's wife's diary there are records of numerous conversations, including one in which Owen explained at length his views on homologies and the vertebrate archetype, and another, apparently at the zoo, in which the two discussed transmutation. There were also pre-breakfast discussions and "after tea muscular fibre and microscope in the drawing room."[15] For many years Darwin continued also to draw on Waterhouse for assistance of various kinds. In looking at the record of Darwin's exchanges with Owen, one senses that after their initial collaboration on the fossil Mammalia from the voyage, Darwin was primarily interested in Owen's general views on comparative anatomy and paleontology. Waterhouse, on the other hand, frequently served as expert consultant on the Coleoptera, rodents, and marsupials. In the mid-1850s, when Darwin was trying to assemble evidence for his interpretation of the phenomenon of aberrant genera, Waterhouse supplied him with lists of aberrant genera of Coleoptera and with his own impressions about what constituted aberrancy and about the geographical ranges of aberrants.[16] A decade earlier, Darwin and Waterhouse discussed at length several issues relating to classification, such as the meanings of "typical," "natural classification," "relationship," "low," "perfect," and so on.[17] As with Owen, Darwin's exchanges with Waterhouse occurred at Down and in London (from 1843 Waterhouse was employed at the British Museum and before that as curator of the Zoological Society). The two carried on a moderately extensive correspondence as well.

Darwin's *Autobiography* conveys the impression that after he moved to Down in 1842 he was almost cut off from converse with his fellow naturalists. "Few persons can have lived a more retired life than we have done," he said:

> Besides short visits to the houses of relations, and occasionally to the seaside or elsewhere, we have gone nowhere. During the first part of our residence we went a little into society, and received a few friends

here; but my health almost always suffered from the excitement, violent shivering and vomiting attacks being thus brought on. I have therefore been compelled for many years to give up all dinner-parties; and this has been somewhat of a deprivation to me, as such parties always put me into high spirits. From the same cause I have been able to invite here very few scientific acquaintances.

This may have been true for the ten or twenty years before it was written. But for the 1840s and 1850s Darwin's correspondence and his notes on conversations about scientific matters indicate a rather fuller scientific life.[18] There were fairly frequent trips to London to use the collections, libraries, and periodical holdings of the British Museum, the Athenaeum, and the various scientific societies, and to attend the scheduled meetings of the societies, as well as to meet individually with Lyell, Owen, Hooker, Waterhouse, H. C. Watson, Hugh Falconer, Edward Forbes, and others. There were also the parties at Down, which might equally well be described as small scientific meetings. In groups or singly, with or without wives, Darwin's "scientific acquaintances" were invited for a few days' visit, during which one main activity was scientific discussion. Hooker has left an account from the visitor's perspective of one of Darwin's methods of gaining access to his guests' knowledge:

The next act in the drama of our lives opens with personal intercourse. This began with an invitation to breakfast with him at his brother's (Erasmus Darwin's) house in Park Street; which was shortly afterwards followed by an invitation to Down to meet a few brother Naturalists. In the short intervals of good health that followed the long illnesses which oftentimes rendered life a burthen to him, between 1844 and 1847, I had many such invitations, and delightful they were. A more hospitable and more attractive home under every point of view could not be imagined – of Society there were most often Dr. Falconer, Edward Forbes, Professor Bell, and Mr. Waterhouse – there were long walks, romps with the children on hands and knees, music that haunts me still. Darwin's own hearty manner, hollow laugh, and thorough enjoyment of home life with friends; strolls with him all together, and interviews with us one by one in his study, to discuss questions in any branch of biological or physical knowledge that we had followed; and which I at any rate always left with the feeling that I had imparted nothing and carried away more than I could stagger under. Latterly, as his health became more seriously affected, I was for days and weeks the only visitor, bringing my work with me and enjoying his society as opportunity offered. It was an established rule that he every day pumped me, as he called it, for half an hour or so after breakfast in his study, when he first brought out a heap of slips with questions botanical, geographical, &c., for me to answer, and concluded by telling me of the progress he had made in his own work, asking my opinion on various points.[19]

Many of the slips of paper Hooker referred to are in the Black Box envelopes, often with a large "Q" (for "Query" or "Question") on them. Here is a set of "Questions for Hooker" which Darwin used during a visit in April 1852:

(1) Instances of genera varying in one quarter of world & not in another quarter . . . Does any one individual species keep pretty constant in one quarter & vary in another, or vary in one particular way in one district & differently in another?
(2) Do the local or widely distributed species vary most . . .
(6) Cases of plants which must *for-ever* self-fecundate themselves; do such occur? as in a plant which fecundates in bud?[20]

Often the answers to such questions are recorded in the numerous notes on conversations that are scattered throughout the Darwin manuscripts. It is not clear from the notes I have seen whether Lyell, Owen, Forbes, or several others ever were "pumped" by Darwin in the manner described by Hooker, although Huxley, Waterhouse, and perhaps Thomas V. Wollaston were. But whichever naturalists were present, there was always conversation on scientific subjects, and usually, it would seem, Darwin managed to learn much that was interesting and useful to him.

When personal meetings were impossible, Darwin made use of the mail to find out what he needed to know. Much of his most important scientific correspondence was with members of this same circle of acquaintances, that with Hooker being particularly extensive and valuable; but he also received considerable aid from naturalists he never or only rarely saw, such as Asa Gray. With Hooker he discussed in letters nearly every conceivable topic relating to geographical distribution, as well as classification, embryology, the meaning of "highness and lowness" in botany and zoology, and much more. From Gray and others he received data on which he based the botanical statistics he used in his work on divergence and variation in large genera. Darwin appears to have been very adept at getting people to help him who were in a position to supply detailed information, technical assistance, and informed professional opinion on a variety of subjects.[21]

But Darwin needed more than this for his projected work. From 1838 on his aim was to reform all of the natural history sciences.[22] This placed on him the burden of reinterpreting what was already known, and for this he required at the very least a good working knowledge of the principal generalizations of his fellow naturalists. With the exception of Owen, and possibly Hooker, none of Darwin's botanical and zoological friends was a major figure in theoretical natural history. They could tell Darwin what others had proposed, but their statements could not be taken as authoritative, nor could Darwin expect them to be interested in or to appreciate or even to be aware of all of the latest innovations in biological science. Necessarily, therefore, he relied very heavily on books and periodical literature, reading "books of all kinds" and "whole series of Journals and Transactions," making abstracts of them or scribbling all over them and, inside the back cover, preparing an index to his scribbling.[23] The fact that Darwin read may seem at first sight so obvious as scarcely to be worth mentioning. Anyone who has looked at two or three books by Darwin or skimmed through the bibliography of *Natural Selection* or read Francis Darwin's "Reminiscences" of his father is well aware that Darwin read widely in the biological literature of his day. Nevertheless, it seems to me that Darwin's reading is not generally appreciated for what it was – namely, a major scientific activity, as intimately bound up with his theoretical endeavors as any of the other work that more readily comes to mind in response to the phrase "Darwin's scientific activities" – such as the *Beagle* voyage, with its collecting and observing; the years of morphological and taxonomic work on barnacles; the pigeon breeding; the experiments on transport of seeds; the grass collecting with the governess; the observations on insects and flowers, earthworms, climbing plants, and the facial expressions of dogs and infants; and the experiments with weeds and grass on the lawn at Down, to mention only some of the more noteworthy. There can be little doubt that reading was more important to Darwin than any of these, with the possible exception of the voyage, without which perhaps none of the rest would have followed.

It may be that it is Darwin's way of speaking of his work that has caused the importance of his reading to be underestimated. By his own account he read primarily for the purpose of collecting "facts."[24] It is the word "facts," which Darwin used rather freely, that is so misleading. Darwin of course did collect and use a great many observations and low-level generalizations that can best be called facts. But he also collected facts of an entirely different order. These are what he sometimes called, and more accurately, "classes of facts," "laws," "rules," and "generalizations." Thus Darwin might have designated as a fact Owen's interpretation of the structure of the *Toxodon* – that it was a rodent having affinities with the pachyderms and Cetacea.[25] This was a "fact" that Darwin delighted in, for the *Toxodon* exhibited just the sort of intermediate structure that fossil forms ought to have if presently existing groups have branched off from common ancestors: "How wonderfully are the different Orders, at the present time so well separated, blended together in different points of the structure of the Toxodon!" he exclaimed.[26] Owen, partly on empirical and partly on theoretical grounds, soon concluded that such combination of characters was a general paleontological law: the more ancient the fossil, the more likely that it would combine features of several now-distinct groups. This too Darwin liked very well, and it figures prominently in the paleontological discussion in the *Origin*.[27]

Two other examples illustrate even more clearly the nature of the "facts" Darwin gleaned from his reading. Unity of type was so far from being a fact for teleologists like Cuvier or Buckland or Charles Bell that they in effect denied that there was any such phenomenon in nature. If vertebrates are all built along similar lines, they said, it is not because there is a law of unity of type, but rather because all are required to perform similar functions. Geoffroy, on the other hand, suggested that unity of type extended throughout the animal kingdom. Most, though not all, of the leading biologists of the 1840s and 1850s accepted the doctrine, but they said its applicability was restricted to members of a common *embranchement*. The doctrine of unity of type was a major, controversial, theoretical statement of mid-nineteenth-century zoology. Yet Darwin on occasion spoke of it as one of the "great facts" in morphology. Similarly, von Baer's law that among the members of each great group of animals the embryos are fundamentally alike in their earliest stages was for Darwin "the grandest of all facts in . . . zoology."[28] In the light of this usage of the word, Darwin's portrayal of himself as a collector of "facts" takes on a new significance.

It is precisely such facts as these last – unity of type, the law of embryonic resemblance, various laws of geological succession and geographic distribution – that form the basis of the chapters in Part II, where they receive an evolutionary explanation. We cannot dismiss these laws as obvious, or as common knowledge that only became significant when Darwin showed how to interpret it, for they were recognized in Darwin's own day – and not the least by Darwin himself – as new and important theoretical developments in natural history. Darwin's explanations are in most cases so direct and simple that we forget that the generalizations he was explaining were themselves the product of considerable effort and creative thought by his contemporaries. In many cases the initial effort of formulating the law was much greater than that required to transform it into a principle of evolutionary biology. The law of unity of type was firmly established only as a result of the morphological and embryological work of Geoffroy, Milne Edwards, von Baer, Owen, and many others. Yet everyone, whether he was a transmutationist or not, understood that unity of type could be explained by the doctrine of descent – that, after all, was one of the principal reasons it was opposed by the teleologists.[29]

Other laws, however, could not as easily be fitted into an evolutionary mold, either because they represented an inadequate understanding of the phenomena or were wrong, from an evolutionary point of view; or because their evolutionary interpretation was not obvious. In such cases Darwin had to decide whether to reject or ignore the law or to attempt to find an explanation for it, and in these instances he was guided by the extent to which the law was approved of by his contemporaries. Sometimes he could identify an "error," as when he wrote in the margin of one book "all this discussion strikes me as unsatisfactory, from struggle with other species, not being more prominent."[30] If a law seemed fantastic or mystical to others besides himself, Darwin might admit that he could not explain it (when in 1854 he began to prepare his species work for publication, he wrote inside the cover of the "Essay of 1844" the instruction "make list of laws which I cannot explain as Quinarism – Forbes' Polarity, Barrandes Colonies &c &c. These shortly discussed w^d be interesting").[31] But usually he was inclined to credit the laws his colleagues proposed, even those that did not lend themselves readily to an evolutionary interpretation.

Darwin believed that his theory was right, and so he believed that every fact or law that was well attested to by competent naturalists should be at least compatible with, if not a direct consequence of, evolution by natural selection. His need to be sure, and to demonstrate to others, that all the facts and laws of natural history could be encompassed by his theory meant that reading was not for him a passive accumulation of information, but was rather an active speculative endeavor. Take for instance the following generalization from Alphonse De Candolle's *Géographique Botanique Raisonée:* "where the vegetation is more varied than in Europe, the habitats are generally smaller, and it is probable that many species are limited to a single locality." Darwin's marginal comment is part of his attempt to explain not only geographical distribution, but also classification, and it is a step toward the principle of divergence: "this bears on few species inhabiting 2 areas, where there are many species. Does it not come to this, that widely extended species break into varieties & these become species with confined ranges. – Anyhow this shows how complicated question it is."[32] This is not Darwin collecting facts. It is Darwin in conversation, we might say, with De Candolle, reacting to De Candolle's ideas and being stimulated by them to speculative efforts of his own. Similarly, a note in the margin of Milne Edwards's *Introduction à la Zoologie Générale* is part of Darwin's attempt to explain embryonic resemblances by his theory that variations are caused by changes in external conditions: "One feels an early embryo more independent of outside world, but why? so less apt to vary – the late-formed parts exposed to sum of influences, & selection w^d not act on embryo."[33] The picture of Darwin that emerges is not of an isolated, independent scientist, but of a scientist constantly interacting with professional colleagues, either in person, by correspondence, or through the medium of the printed page. This simple fact, if I may call it such, determined the procedure Darwin most frequently followed in the Part II sections of his work.

Darwin's statements about his method in Part II are exceedingly numerous. The transmutation notebooks, the "Sketch" and "Essay," the Black Box materials, the *Origin,* and Darwin's correspondence abound with them. Sometimes, as in the heading for Part II in the "Essay," he said simply that he was presenting the "evidence favourable and opposed" to the idea of common descent.[34] In the *Origin*, similarly, he said at one point that the question to be answered was whether the facts and rules relating to geological succession accord better with the idea of immutability or with descent and natural selection.[35] On several occasions he explained further that if the "scheme" on which all organisms have been produced should prove on examination to be the same as would result from common descent, then it is "highly improbable" that they have been separately created. Another favorite argument was that in light of the theory of descent terms such as affinity and unity of type "cease to be metaphorical expressions and become intelligible facts."[36] Around the time of the publication of the *Origin* Darwin in his correspondence insisted particularly on one of his earliest ideas about what would constitute evidence for his theory: "change of species cannot be directly proved, and . . . the doctrine must sink or swim according as it groups and explains phenomena." A hypothesis, he said, "is *developed* into a theory solely by explaining an ample lot of facts."[37] If we recall what Darwin meant by "facts" we can see from these various statements that what he was doing above all was grappling with the biology of his contemporaries, trying to demonstrate its compatibility with his theory.

This appears most clearly in Darwin's manner of presenting his proofs in Part II. Although Darwin adopted no uniform style of argument, two general approaches can be discerned. Sometimes he deduced consequences from his theory and then compared them with the facts. His discussion in the "Essay" of the absence of intermediate forms is a case of this kind. First he explained what ought to be expected according to his theory:

As I suppose that species have been formed in an analogous manner with the varieties of the domesticated animals and plants, so must there have existed intermediate forms between all the species of the same group, not differing more than recognised varieties differ. It must not be supposed necessary that there should have existed forms exactly intermediate in character between any two species of a genus, or even between any two varieties of a species; but it is necessary that there should have existed every intermediate form between the one species or variety [and] the common parent, and likewise between the second species or variety, and this same common parent.

He then asked "what evidence is there of a number of intermediate forms having existed, making a passage in the above sense, between the species of the same groups?" As the answer in this case was very little, Darwin had to add a further step to the argument in order to explain the discrepancy between expectation and result, which he did by reference to the imperfection of the geological record.[38]

Much more often he employed another form of argument, nearly the reverse of this: First he stated the prominent laws or facts in a subdiscipline of natural history, and then he posed the question of whether they could be explained by his theory. In Chapter X of the *Origin,* for instance, Darwin listed the "several facts and rules relating to the geological succession of organic beings" – laws proposed by Owen, Agassiz, F. J. Pictet, and others – and then argued that they would result from descent and natural selection. In discussing classification he observed that naturalists find most valuable as taxonomic characters not those of high physiological importance, but those that are most constant; and this he was able to explain by his theory, as well as the importance of embryological characters. He briefly summarized the principal generalizations in morphology and then declared that their explanation "is manifest" on the theory of natural selection. In the section on embryology he stated the law of embryonic resemblance, and several others, and then said that all could be explained by descent with modification.[39] This form of argument, which predominates in Part II, was a convenient way for Darwin to persuade others of the truth of his ideas. But more than this, it was a direct consequence of the nature of his project – the reform of natural history – and it reflects his actual procedure in carrying out the reform. In reading and conversation he encountered the leading generalizations of natural history. He then attempted to formulate evolutionary explanations for them. In presenting his case he followed the same order: He began with natural history as he found it, and then he proceeded to recast it. The generalizations of Darwin's contemporaries, in other words, were often the stimuli that induced him to formulate new hypotheses that could later be tested against "facts" of various kinds.[40]

Darwin's procedure in Part II inevitably led to revisions of his theory because in difficult cases explaining the generalizations of others required him to add to his theory, to extend it by inventing new subordinate principles and hypotheses and by developing out of the principles he was already using implications he had not previously considered. In this regard one consequence of Darwin's perspective, from his position as a transmutationist, is particularly important. The laws his contemporaries proposed, or rather those that he accepted, necessarily appeared to him as tendencies in the evolutionary process.[41] A statement by a nontransmutationist about what phenomena exist in nature was for Darwin a challenge to prove that transmutation can be expected to produce those phenomena. The law of unity of type indicated to Darwin that there is a tendency for basic structural arrangements of animals to remain constant. If vertebrate embryos at an early state are all alike, Darwin supposed that there must be a tendency for them to be left unaltered by changes in the mature animal. When embryologists, paleontologists, and taxonomists postulated laws of branching or divergence, Darwin concluded that there must be a tendency to diverge built into the process of transmutation. As a result of this attitude Darwin was frequently faced with the task of showing why transmutation involved tendencies that his own speculations had not led him to expect. An instance of this is one of Darwin's earliest attempts at recasting a contemporary theory, his confrontation with MacLeay's quinary system of classification.

An example of Darwin's method: his explanations of quinarianism

In 1841 Darwin complained that there were so many disputes about "affinity, linear, circular arrangement, &c," that the whole of classification was a mass of "vexed questions."[42] A major source of difficulty was that in post-Waterloo England the search for the natural system took a peculiar turn, producing schemes of classification that now appear to be fanciful contrivances but which Darwin for a time took seriously, in part because he respected several of their authors, in part because their popularity among zoologists meant that they could not simply be dismissed without argument. The work which served as the principal model and inspiration for these schemes was William Sharp MacLeay's *Horae Entomologicae,* published in 1819–21 in two parts, the first being an essay on the Linnean genus *Scarabaeus,* the second, and much longer, a general theory of classification.[43]

MacLeay believed that the natural system, which he called "the immediate aim, of every modern observation in natural history," represented "the plan by which the Deity regulated the creation" and that in attempting to discover it naturalists might obtain "a view of the universe as it was originally designed."[44] The supposition that the natural system was God's plan was widely shared. What particularly distinguishes MacLeay's theory is the kind of order, the formal and numerical regularity, he professed to find in nature. MacLeay subscribed wholeheartedly to the Linnean maxim that there is no *saltus* in nature. The "chasms" that appear to separate some groups from others may be due to our not having discovered all the organisms that inhabit the remote corners of the world; or to the extinction of many forms; or even, he suggested, to the fact that some members of the series have never been created. They do not indicate that the creator has at some points left the chain imperfect.[45] "If we knew *all* the species of the creation, their number would be infinite, or in other words, . . . they would pass into each other by infinitely small differences."[46] MacLeay admitted that the progress of natural history had made untenable such conceptions as Charles Bonnet's of a simple chain of organisms. But he denied that this meant there is no continuity in nature, and he censured those who, like Cuvier, abandoned altogether the idea of the animal series.[47] MacLeay credited Cuvier, along with Lamarck, with founding the modern study of zoology. But while Cuvier was undoubtedly the best comparative anatomist in the world, his efforts to find the natural system were, in MacLeay's opinion, far inferior to those of Lamarck, who never lost faith in the animal series. Lamarck's "merits in Natural History," he said, "bear much the same relation to those of M. Cuvier, that the world has been commonly accustomed to institute between the calculations of the theoretical and the observations of the practical astronomer."[48]

MacLeay did not approve of Lamarck's theory of transmutation, which he thought Lamarck's own work in classification amply refuted.[49] But in Lamarck's attempt at a natural arrangement of the invertebrates MacLeay saw the first, though still distant, glimpse of the system he himself revealed more fully in his *Horae Entomologicae*. Lamarck in his *Histoire Naturelle des Animaux sans Vertèbres* constructed two branching series of organisms, each beginning among the lowest invertebrates. Between their ends he placed the vertebrates, as if to suggest that the molluscs, on the one hand, and the arachnids, crustaceans, and cirripedes, on the other, branched on somehow toward the most perfect animals.[50] In such an arrangement MacLeay saw the possibility of expressing both the continuity he believed to exist in nature and the complexity revealed by modern investigations of internal structure. "Nature," he said, "appeared to me to have branched out in the animal kingdom . . . in a most beautiful and regular though intricate manner, that might be compared to those zoophytes which ramify in every direction, but of which the extreme fibres form by their connexion the most delicate circular reticulations."[51]

Although the idea of ramifications in every direction may suggest disorder and confusion, MacLeay believed in fact that the whole assemblage of "circular reticulations" was, when properly understood, perfectly regular. The animal kingdom he conceived to be a great circle composed of the principal animal forms, and he thought the vegetable kingdom formed a similar circle, the two circles meeting at the points occupied by the lowest members of each group. He supposed the circles ran parallel to each other in the following fashion: Each has five classes, and the classes of one correspond to those of the other. Put in another way (which MacLeay also employed), each circle is a series of affinity that returns on itself. When the affinities of the animal classes are followed, they are found to lead from A_1 to A_2 to A_3 to A_4 to A_5 and from A_5 back to A_1. The same is true among the plants, whose series of affinity is parallel to that of the animals, P_1 corresponding to A_1 and so forth. [52] This same system of arrangement occurs at every level. The circle of the animal kingdom consists of the classes Acrita, Radiata, Annulosa (Articulata of Cuvier), Vertebrata, and Mollusca. Each of these five classes consists of five orders, whose affinities also run in a circle. Among the vertebrates, for instance, the series proceeds fish, amphibians, reptiles, birds, and mammals, the mammals returning to the fish via the Cetacea. In each order there are five tribes, again circular, and so on down the scale.[53] At each level there are correspondences between the members of the five circles. Thus, each vertebrate order corresponds to one order of Acrita, one of Radiata, one of Annulosa, and one of Mollusca.

MacLeay reconciled the existence of separate circles with his faith in continuity by having the circles touch each other, or "inosculate." In each circle of five groups, two groups are "inosculant," serving to join their circle to the two neighboring circles. The continuity is further strengthened by what MacLeay called "osculant" types. These small groups, usually having only a few members according to MacLeay, were the peculiarities of the animal kingdom, the forms that naturalists commonly referred to as aberrant, such as the platypus, *Ornithorhynchus*.[54] Frequently these seemed to unite in a single animal some prominent characteristics of two distinct groups. MacLeay placed them between circles, rather than in them. The tunicates, for instance, are osculant between his Acrita and Mollusca, the cephalopods between Mollusca and Vertebrata, and the cirripedes between Radiata and Annulosa. "Though I could find many chasms in the chain," he said, "no where, after an accurate examination, was it certain that any anomalous interruptions occurred. Nay, the singularities of the animated part of the creation which had hitherto appeared so extraordinary to naturalists, as serving only to defy all arrangement, were here usually the very links required in order to arrive at connexion."[55]

MacLeay's joining molluscs to vertebrates by having cephalopods blend into reptiles is reminiscent of the more peculiar links in Bonnet's chain of being. And his path from mammals to fish by way of Cetacea, as well as the suggestion that the affinity of turtles and birds is demonstrated by the hawksbill turtle, indicates that he was less able than many of his contemporaries to distinguish between true relationships and merely adaptive resemblances.[56] Yet in MacLeay's eyes the great advantage of his system of arrangement was precisely that it gave expression to, and honored the distinction between, two different kinds of relationship, affinity and analogy, which he defined as follows:

A natural series of affinity is such as, taking the majority of characters for our guide, shall be found uninterrupted by any thing known, although possibly broken by chasms occasioned by the absence of things unknown.[57]

Relations of analogy consist in a correspondence between certain insulated parts of the organization of two animals which differ in their general structure.[58]

The series of affinity are the chains of relationships that form circles, connect groups. They join together to form the branching but continuous animal series. Analogies are the correspondences between the members of different parallel series.

MacLeay's contemporaries were generally agreed that he had done good service by distinguishing clearly between affinity and analogy, and it does seem that his work and the discussions it prompted made naturalists more conscious of the importance of discriminating carefully between the two kinds of relationship.[59] But there are two points about MacLeay's distinction that should be noted. He did not claim to be the first to distinguish between affinity and analogy in natural history, but rather to have discovered "the *nature* of the difference" between them. And the "*nature* of the difference," as MacLeay explained it, was not at all what the comparative anatomists and taxonomists of his day (and since) had in mind.[60] For them the important distinction was that between, in Owen's words, "essential" and "adaptive" relationships or characters.[61] This is a distinction that is implicit in the work of Geoffroy, but it was Owen who first gave reasonably clear definitions and adhered to them consistently: "Analogue. A part or organ in one animal which has the same function as another part or organ in a different animal." "Homologue . . . The same organ in different animals under every variety of form and function."[62] In evolutionary terms this is the distinction between the results of common descent and of convergence. According to this way of looking at organisms, the taxonomist's business is to discern relationships in essential characters; merely adaptive resemblances, far from being part of the natural system, are likely to mislead the naturalist and confuse him in his efforts to arrange organisms as naturally as possible. But in MacLeay's definitions one finds no reference to adaptive or functional similarities as distinct from essential or formal. For MacLeay, who in this was truly a follower of Cuvier, all characters are adaptive. The difference between affinity and analogy is simply that the first is determined by agreement in the majority of characters, while the second is a similarity in a few particulars. When the analogies between two forms become very numerous, they "compose" an affinity. In this case two of the analogous forms in neighboring circles are also the inosculant forms that connect the circles (by affinity). Furthermore, relations of affinity and of analogy are equally parts of the scheme. They are two relationships both of which are to be expressed in a "natural" arrangement: "The natural system ought to express all the relations which exist between the various objects of our study, and . . . if any of these are left unrepresented in our arrangement, there must be some latent error, in the formation of our groupes . . . the natural arrangement . . . is to be attained only by the expression of every affinity, and every analogy that can be detected."[63]

The fact that analogies are as much a part of the scheme as affinities meant for MacLeay that instead of causing confusion, analogies are positively useful to the naturalist. If he has trouble determining the order of affinities in a particular group, comparison with a neighboring series in which the affinities are well established may show him the solution. It is only necessary for him to see with which forms in the other series his questionable forms are analogous, and their positions will be clear. And when the taxonomist has arranged his material according to its relations of affinity, the existence of proper analogies with other series proves his arrangement is correct, that is, agrees with the creator's plan. "It is fortunate," he said, "that the naturalist cannot have a more admirable test of his accuracy, or a stronger rein on his fancy, than this parallelism of analogous groups in contiguous series of affinity. Thus, although a solitary resemblance may mislead, it is clear that when we find several of such resemblances to keep parallel to each other in contiguous series, we may reckon upon their having some more solid foundation than our own fancy."[64]

MacLeay claimed to have been criticized for arguing that there is no single chain of being or order of progression in nature, a notion opposed to the dictates of revelation, as well as of reason. He responded that on the contrary, so far was his plan from militating against the doctrines of revealed religion, that it depended on these "as some of its very best supports." By this he presumably meant that it is faith in God that gives confidence in order and plan. Conversely, of course, he held that finding such order in nature as he did was proof of design, for the numerical and even geometrical regularity of his system could not be contrived by "the utmost human ingenuity." Modestly, he declared that his system differed from all others because they – being artificial – merely simplify the means of acquiring knowledge, while his, insofar as it is correct, reveals one of the creator's "stupendous works."[65] But MacLeay did not rest the argument for his theory only on faith or design. He insisted that he had never adopted a new theory until it "had almost lost the right to the name, and had become a mass of facts, observed indeed by others, but now for the first time arranged so as to form one regular whole."[66] In addition to the facts of entomology, as observed by himself and others, which, as he thought, pointed to classification in circles of five, MacLeay noted that the elder De Candolle, quite independently, detected a quinary arrangement in the Cruciferae and that the botanist Elias Fries found the distribution of fungi to be quinary. "I cannot help rejoicing," he said, "that the strength of this beautiful theory should be so completely brought home to the conviction of every mind, as it must be, by observing the manner in which different persons have respectively stumbled upon it in totally distinct departments of the creation."[67]

It should not be surprising if such a coincidence made an impression on MacLeay's readers; nor is it to be wondered at that in a nation renowned for its "peculiar love of regarding Nature from a theological point of view" MacLeay attracted a band of disciples who worked actively to prove and complete his theory.[68] The Zoological Club of the Linnean Society and the Club's *Zoological Journal* (1825–30) devoted much time and space to discussions of quinarianism, regular systems of analogies, and circular arrangements. Others somewhat outside this group contributed to these discussions as well, such as Darwin's friend Waterhouse.[69] Besides the many now nearly forgotten naturalists like Henry Thomas Colebrooke and Nicholas Vigors who were quinarians, a large number of leading British biologists of the mid-nineteenth century were at some point in their careers serious students of MacLeay's ideas. Richard Owen entered MacLeay's address in one of his early notebooks; he copied, or had copied for him, page after page of the *Horae Entomologicae;* and he cribbed extensively from MacLeay's book in preparing his first course of Hunterian Lectures for the Royal College of Surgeons[70]. John Edward Gray, later Keeper of Zoology at the British Museum, followed the quinary system in his arrangement of reptiles in 1837.[71] William B. Carpenter published MacLeay's classification of the animal kingdom in an early edition of his *Comparative Physiology,* and Robert Chambers, who drew heavily on Carpenter, adopted the quinary system in *Vestiges of the Natural History of Creation.*[72] Thomas Henry Huxley in his youthful search for laws of harmony in the organic world found much to admire in the ideas of MacLeay, whom he met in Sydney.[73] And the evidence of the transmutation notebooks makes it clear that Darwin too must be included among those who thought for a time that there was something to quinarianism, though the something was not what Huxley or MacLeay saw in it.

Darwin knew MacLeay (MacLeay was one of the zoologists who most strongly urged Darwin to publish the zoology of the *Beagle* voyage); he participated in discussions with him, probably at the Zoological Society; and he studied carefully the *Horae Entomologicae* and MacLeay's most important papers on classification, as well as much of the literature generated by the widespread interest in quinarianism.[74] In the transmutation notebooks alone I have found over eighty pages on which Darwin took note of or responded to ideas associated with MacLeay's system, and his reading notes, correspondence, and folders of notes on Part II subjects contain many more. Until recently Darwin scholars have routinely treated MacLeay's system simply as an obstacle to Darwin, an absurd theory which he had to prove false in order to establish the theory of natural selection.[75] Darwin did eventually come to see most of MacLeay's system as absurd. But what is most interesting about his numerous early references to it is that they do not for the most part represent instances in which Darwin was developing arguments to show that MacLeay was wrong. His initial assumption was rather that MacLeay was more or less right. That is to say, Darwin assumed that there was a basis in nature for MacLeay's claims to have found that natural groups usually contain five subgroups and that there are analogical resemblances running through all the groups of the animal kingdom.[76] For a time Darwin treated these laws of MacLeay, along with the existence of osculant forms, as the results of tendencies in the evolutionary process, and he attempted to explain why analogous resemblances and groups of five are apparently regular, generally occurring consequences of transmutation.

Darwin's explanations of osculant forms and of the circularity of groups are relatively straightforward. He suggested that circularity might be due to extinction. When the lower branches of the tree of life die, he said, early in the first notebook, the upper branches that remain, viewed from above, might "appear like circles."[77] Later, when he had come to see much of MacLeay's system as the invention of its author, rather than the creation of nature, he concluded that circles were a main source of the quinarian's self-deception. He argued in a letter to Waterhouse, who arranged the mammals in circular groups, that circles of equal size inevitably suggested groups of equal value when in fact decisions about the value of groups were arbitrary, depending on the number of forms they contained, habitat, convenience, and so forth; and he concluded, "as for your wicked circles, I wish they were all d – d together."[78] But as long as circles were a prominent feature in the taxonomic literature, Darwin was obliged to provide a plausible explanation for them within the framework of his theory. Such a one is sketched in the third notebook: "Argument for circularity of groups. When a group of species is made, father probably will be dead – hence there is no central radiating point, all united . . . now what is group without centre but circle, two or three lines deep."[79] Osculant forms Darwin decided after some hesitation were "living fossils," remnants of a "parent intermediate race."[80] He concluded that they would probably be small and well defined, as MacLeay said they were, because of the extinction of their relatives.

MacLeay's regular system of analogies, and the hypothesis of equal numbers in each group on which the system depended, presented a rather more difficult problem. Darwin from the outset supposed that in a truly natural system organisms are grouped according to their degree of genetic relationship[81] If so, the best arrangement would be that based on characters inherited from common ancestors. "In this view," Darwin wrote Waterhouse, "all relations of analogy &c &c consist of those resemblances between two forms, which they do not owe to having inherited it, from a common stock."[82] MacLeay, on the other hand, said that analogy is as much a part of the natural system as affinity. Nature is observed to present groups with equal numbers of forms and with regular analogies connecting their members. For Darwin to account for this apparent taxonomical law, he had to devise new explanatory principles in addition to common ancestry. One of these appears in his first discussion of the tree of life: "Would there not be a triple branching in the tree of life owing to three elements – air, land and water, and the endeavour of each typical class to extend his domain into the other domains and subdivision[s] [gives] three more [and thus a] double arrangement – If each main stem of the tree is adapted for these three elements, there will be certainly points of affinity in each branch."[83] Here the existence of three main kinds of environment and a presumed tendency of forms primarily adapted for one to give rise to forms adapted for the other two are called on to give the tree of life a somewhat regular shape and thus to explain analogies and numerical constancy. The number produced by this hypothesis would seem to be three or nine, but Darwin's further statements indicate that he thought he could explain the prevalence of the number five, on which MacLeay insisted:

> The Creator has made tribes of animals, adapted preeminently for each element, but it seems law that such tribes, as far as compatible with such structure, are in minor degree adapted for other elements. Every part would probably be not complete, if birds were fitted solely for air and fishes for water – If my idea of origin of Quinarian system is true, it will not occur in plants, which are in far larger proportion terrestrial – if in any in the Cryptogamic flora – but not atmospheric type hence probably only four, is this not Fries rule, what subject has Mr Newman the (7) man studied.

The endeavor of each tribe to extend its domain is now a "law," the first of many laws and principles Darwin formulated in order to enable his theory to account for the generalizations of contemporary natural history. Unfortunately he did not explain how exactly this particular law produced the desired results. An entry made at the end of the first notebook some months later is an instruction to "work out Quinary system according to three elements," but as far as I know no notes remain in which Darwin did so.[84]

The passage just quoted indicates that Darwin thought the number might vary in different groups; five might hold among animals, but not among plants. In this Darwin was beginning to depart from MacLeay, and within a few months he concluded that numerical regularity was not to be expected in nature. Chances are that every family will have some forms analogous to members of other groups, he said, but "number five in each group absurd." Darwin at this point (the spring of 1838) still believed in a plan of creation, but his idea of it was nothing like MacLeay's. He imagined a continual round of geological changes; alterations in climate; division and reunion of land masses; and migration, transportation, variation, and extinction among organisms. He had decided that his theory of "breeding in irregular trees and extinction of forms," as he expressed it, was not compatible with the regularity of quinarianism. More and more Darwin insisted that this tenet of MacLeay's system was necessarily and absurdly erroneous. "Anyone may believe anything in such rigmarole about analogies & numbers," he said.[85]

At the same time, Darwin continued his efforts to explain the apparently universal law that every group includes forms analogous to the forms in other groups. The basis for most of his speculations on the subject was, throughout the notebooks, the three-element hypothesis. Forms are analogous because they are adapted for life in the same environment: "similar habits produce similar structure."[86] But while this might well explain each individual case of analogy, alone it could not account for the generality of the phenomenon, the fact that analogous forms are numerous, seemingly occurring in every group. This Darwin at first said was caused by the "attempt in each dominant structure to accommodate itself to as many situations as possible." Subsequently he reformulated this hypothesis in less anthropomorphic terms. Analogies will be formed wherever the "scheme" of nature is not "filled up." If an animal comes to an island to which it is not perfectly adapted and where many places are open, it may, from adopting new habits, be so changed as in some respects to resemble another group more nearly than its own. The chances are, Darwin argued, that this will have happened in every family, and so all will probably have their terrestrial, aerial, and aquatic types. Some of the best-known passages in the notebooks are speculations along these lines. Darwin's Lamarckian explanation of hypothetical changes in the habits and form of a jaguar is to illustrate the production of an aquatic type in a group of land animals: "for instance, fish being excessively abundant & tempting the Jaguar to use its feet much in swimming, & every developement giving greater vigor to the parent tending so [to?] produce effect on offspring." The suggestion that "monkey would breed (if mankind destroyed) some intellectual being though not MAN" serves a similar purpose. The destruction of man would leave empty a place in the "scheme." The new forms that would fill it would be analogous to the now extinct human species, but their affinity, which is determined by descent, would be to the monkeys.[87]

By the spring of 1838 Darwin had arrived at a fairly clear

understanding of convergent evolution and of the difficulties it might cause in distinguishing between affinity and analogy:

> Relations of analogy being those last obtained [are] less firmly fixed & therefore most subject to change . . . The same characters which are analogical in a genus with respect to rest of its family as in ground cuckoos, is affinity with respect to species of each other, because we suppose all descended from same – but if two original species, each became ground, then the relation of all the ground cuckoos would not be affinity, but the truth would never be discovered.[88]

While this explanation of convergence is notable as a contribution to evolutionary theory which has proved to be of continuing interest, I want particularly to emphasize what led Darwin to it. Darwin was not simply trying to account for a few odd facts – those cases in nature in which two animals of different basic structure have come to resemble each other more or less closely in outward form. He was trying to show how his theory could explain one of the principal generalizations associated with MacLeay's peculiar system of classification. Darwin denied that nature was as regular as MacLeay imagined. But he accepted from MacLeay (and others) the dictum that relations of analogy are a general, not an isolated, phenomenon.

It was the presumably general occurrence of analogies that imposed on Darwin the necessity of inventing new principles to enable his theory to account for them.[89] If descent is the only cause of similarities, one could not expect to find more than a few chance resemblances among groups that are not descended from the same stock. But naturalists found analogous resemblances to be common. Since heredity could not be the cause of them, Darwin had to add to his theory to show why they should exist among organisms produced by "breeding in irregular trees" rather than by God's geometrical design. He did so by introducing the three-element hypothesis into his theory and by arguing that it is probable that the tendency of living beings to "fill up" the "scheme" of nature will cause every family to present analogies with other groups.[90]

Darwin continued to speculate on the causes of analogous resemblances, and in the *Origin* he could not resist mentioning his three-element hypothesis.[91] But developments in the early 1840s relieved him to a considerable extent of the need to worry about quinarianism as such. Darwin initially thought there might be something to MacLeay's circles and numbers, but even after he had concluded it was all a "rigmarole," he was required to work at explaining it because it was so widely accepted. As long as large numbers of naturalists believed in numerically constant groups and regular correspondences among them, Darwin, if his theory was to succeed, had to show why evolution might be expected to produce these appearances.[92] By the mid-1840s however, quinarianism had been rejected by most of the better British zoologists. The culmination of this movement came at the meeting of the British Association in 1843, when Owen, Hugh Strickland, Edward Forbes, Edwin Lankester, and others condemned quinary and circular systems.[93] For Darwin this meant that he need not discuss the subject in his work. He wrote in March 1844 while he was drafting the "Essay of 1844," "It is an hallucination, to suppose that Quinarianism can be explained by air, earth & water – if it were only analogical resemblances it might be so – But as botanists make number of groups & no aerial plants – better to omit whole subject – "[94] Robert Chambers's publication of the *Vestiges of the Natural History of Creation* later the same year caused Darwin to reconsider his resolution to ignore MacLeay. "After the 'Vestiges,'" he said, "I see it will be necessary to advert to Quinary System, because he brings it in to show that Lamarck's willing (& consequently my selection) must be erroneous." Now, however, he believed he could deal with it simply by pointing out that "few English, *sound* anatomical naturalists" accept it, and "hardly any foreign."[95] When he first developed his three-element hypothesis in the transmutation notebooks, it was in order to explain why nature was more or less as MacLeay found it. But when he mentioned it in the *Origin*, it was to show how naturalists had been misled into developing the quinary and other similar systems. The contrast reflects a parallel shift in Darwin's views and those of his professional colleagues.

Conclusion

Gavin de Beer, perhaps better than he knew, epitomized the relationship of Darwin's work to the biology then current when he wrote, "While Darwin was always on the look-out for facts, what he most hoped for in the works of his predecessors and contemporaries was ideas."[96] The material on which Darwin's intellectual labor was most consistently and profitably expended was not raw facts but the generalizations of the leading naturalists of his day. His efforts to make a convincing argument for evolution required his accepting, rather than rejecting, most of the work of his contemporaries, for his goal was to show that their science made more sense when interpreted in the light of transmutation than of the conventional idea of independent creation. To be sure, Darwin saw their science as pre-Darwinian, for he believed that his views had the potential to revolutionize biology. But he did not automatically assume that because it was pre-Darwinian it was fundamentally unsound. He did not suppose that he knew more than his colleagues who were experts in their fields or that his powers of judgment and generalization were greater than theirs. On the contrary, he tended to accept their laws even when, like quinarianism, they were not obviously favorable to his theory. MacLeay's system was only one set of laws Darwin struggled with, however, and it was by no means the most important. It might be called utterly pre-Darwinian, in that Darwin as well as his professional colleagues rejected it, so that it forms no part of his evolutionary synthesis of mid-nineteenth-century biology. The situation is very different in the case of the principal generalizations of mainstream natural history in the period. The work of the post-Cuvier generation of naturalists provided Darwin with much of the best evidence on which his synthesis was based. It also posed problems for him which stimulated some of his most significant theoretical work in the 1840s and 1850s.

5

Natural history after Cuvier: the branching conception of nature

The rejection of the strictly teleological method of Cuvier and British natural theology that I described earlier was accompanied by the establishment of a new general conception of the structure and relations of organisms, one that Darwin worked diligently and on the whole successfully to incorporate into his evolutionary biology. This new conception was in the strict sense of the word a synthesis of the opposing views of Cuvier and the teleologists on the one hand and Geoffroy and the German morphologists on the other. Cuvier's functional approach stressed the differences, indeed the lack of any fundamental relationships, among organisms. Each being was to be considered as a functioning whole whose structure satisfies its own peculiar conditions of existence. Resemblances among organisms, even such basic correspondences in structure as those that caused the naturalist to group all vertebrates together, occurred only as a result of similarities in their conditions of existence. While Cuvier's science emphasized differences, Geoffroy's insisted on fundamental structural similarities, a unity of type extending throughout the animal kingdom. But even as the antagonism between Geoffroy and Cuvier was increasing in the years before their debate in 1830, a new generation of biologists, of which Darwin was a younger member, was beginning to create a synthesis that by the 1850s became the most generally accepted view of the organic world.

Although Cuvier appeared to most observers to have won the debate in the Academy of Sciences, the new synthesis was in fact much closer to Geoffroy's views. Its cornerstone was the law of unity of type, which Cuvier denied.[1] But while as a general method in comparative anatomy Cuvier's teleology was rejected in favor of Geoffroy's approach, some of the functional viewpoint's strongest strictures against Geoffroy were accepted. As a result, a more limited and more rigorously defined law of unity of type was established. Most importantly, the chief architects of the new synthesis agreed with Cuvier that there are several types of structure, not one, and that there are no transitions, ideal or actual, from one type to another, except among the very lowest members of the several great groups of animals. Their principal concern was to trace and reduce to order the similarities among the various representatives of each type; but following Cuvier, they acknowledged the importance of studying function and of recognizing the diversity of adaptations within each type.

By combining unity and diversity in a single scheme, the morphologists of the second quarter of the century produced a branching conception of organic nature, a developmental conception in which diversity was seen as proceeding out of an initial unity. The novelty of the branching conception lay not in the mere idea of development, nor in the image of the branching tree, which its proponents commonly employed. Both can be found in the work of their predecessors – Pallas, Oken, Lamarck, Geoffroy, and others. What made it a new and useful conception in natural history was the way in which branching and development were defined and their limits carefully stated. It was supposed that there was a common starting point for this branching development, whether one was discussing the real development of embryos or the ideal development expressed both in the classification of adults and in the series of fossil animals (these morphologists were usually not transmutationists).[2] With respect to the animal kingdom as a whole, this common point was either the germ, whose simple form was similar in all animals, or the simplest infusorian, which closely resembled the germ. For the major divisions of the animal kingdom – the Radiates, Molluscs, Articulates, and Vertebrates – the common starting points were their respective archetypes; the articulate and vertebrate archetypes in particular were delineated with care and in detail (it was generally doubted whether there was a single archetype for the Radiates). The direction of development was conceived to be not simply upward, nor a linear movement toward man; rather, it was from the homogeneous germ or the unspecialized archetype toward the multitude of diverse and specialized animal forms. The definition of development, and hence of progress, was increasing differentiation and specialization. By its nature, this sort of development –

divergence from common starting points – is limited: Organisms or groups of organisms whose development begins from one point, say the articulate archetype, can never develop into others, such as vertebrates, whose starting point is different. On the other hand, development may be traced backward to a common point of unity from which both articulate and vertebrate and all other types diverge.[3]

The branching conception was naturally attractive to transmutationists, and all of the major theorists of transmutation in the mid-nineteenth century made use of it.[4] A substantial portion of the facts and laws in morphology, embryology, classification, and paleontology that Darwin translated into evolutionary terms in Part II of his work on species were furnished by the branching conception. He found that in most cases they agreed remarkably well with his own views. But even when they did not, Darwin could not ignore them, for the branching conception was very widely accepted by the 1850s, and it numbered among its principal authors some of the most respected biologists of the period, including von Baer, Owen, and Milne Edwards.

Von Baer's embryology

The idea of an archetype, the ideal form on which all the members of any group of organisms are modeled, is the foundation of the branching conception.[5] In 1828 von Baer demonstrated the utility of the idea in natural history by making it the basis of his interpretation of the facts of embryology and by using his observations on individual development to establish workable limits for its application. This is not to say that it had not been employed previously. Students of articulate and vertebrate morphology had already gone far toward determining the most important homologies in their respective groups and toward suggesting common structural plans. But either they tended to avoid large generalizations or they attempted in their general discussions to describe the one archetype that served as the basic plan of all animals, which caused their representations to be so vague as to give little help in understanding either embryonic development, paleontology, or the structural relationships among organisms.[6] Von Baer argued persuasively that each of Cuvier's four *embranchements* has a different archetype; and he effectively

Karl Ernst von Baer.

used a model of the ideal vertebrate to elucidate the process of vertebrate development.[7]

Of the several places in which von Baer discussed the archetypes of the four *embranchements,* the most important is the "Fifth Scholium" of volume 1 of his embryological treatise *Ueber Entwickelungsgeschichte der Thiere.*[8] The subject of the "Fifth Scholium" is the doctrine of recapitulation, which von Baer undertook to

118

refute. Von Baer's opposition to recapitulation had its roots in his dissatisfaction with the linear view of nature on which recapitulation at that time depended.[9] As a result of the increased knowledge of internal structure of animals, and of Cuvier's criticism of the chain of being, virtually no naturalist of any standing in the 1820s arranged animals in a simple series.[10] But adherents of the doctrine of recapitulation retained in their theory the notion of a single scale of organic perfection, and they used the doctrine as a means of establishing the unity of type of all animals. They supposed that the developing embryos of higher animals pass through the adult forms of animals lower on the scale. Oken, for instance, said there is no linear series. Instead, the different classes of animals are parallel to one another in such a way that the highest form of one class stands above the lowest form of the next higher class. But he imagined, nonetheless, a single scale of organic perfection, defined by the addition of organs in a determinate sequence: The infusorian is a simple stomach; the coral obtains an intestine; the acaleph has, in addition, a vascular system; and so on, up to man, who has all the organs that are present in the animal kingdom.[11] Von Baer denied that any such scale exists, and he based his criticism of recapitulation on an analysis of animal form that ruled out any possibility of a linear sequence.

The heart of von Baer's analysis is the distinction he made between "grade of development" and "type of organization." He defined the grade of development of an animal as

> the greater or less heterogeneity of its elementary parts of the separate divisions of a complex apparatus; in a word, in its greater histological and morphological differentiation. The more homogeneous the whole mass of the body is, so much the lower is the grade of its development. The grade is higher when nerves and muscles, blood and cell-substance, are sharply distinguished.

The mistake of the recapitulationists, he said, was to recognize only this one of the two fundamental relations among animals. Their single scale of perfection took into account only differences in grade of development, and so they could imagine that the embryo of the higher animals could pass successively through all the lower grades. But animals differ also in type of organization. "Type" von Baer defined as "the relative position of the organic elements and of the organs." All the animals of a given type are

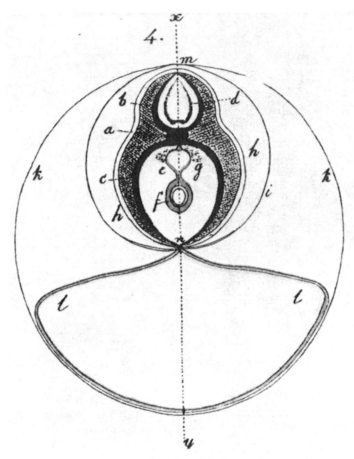

Von Baer's representation of the typical vertebrate embryo, cross section. (*Ueber Entwickelungsgeschichte der Thiere*, I [1828], plate 3.)

modifications of an archetype "in which this relative position is especially characterized." Von Baer described four of these, one each for the radiates, articulates, molluscs, and vertebrates. The type, he emphasized, is totally different from the grade of development, and, conversely, the same grade of development may be attained in many types." If this is so, then a linear arrangement of animals in order of perfection is impossible.[12]

Von Baer made this analysis of the two kinds of organic relations the foundation for his interpretation of the process of embryological development. In the course of development the embryo undergoes a gradual differentiation of tissues and organs. The difference between the embryo in its early and late stages of development is precisely the same as that between the lower and higher grades of development that may be observed among adult animals. "The fundamental mass of which the embryo consists agrees with the mass of the body of the simplest animals." From this it follows, von Baer said, that there must be correspondences between the embryos of the higher animals and the adults of the lower. But there is no recapitulation, for these correspondences are in grade development only. Von Baer's observations on the development of the chick indicated to him that embryos, as much as adults, are characterized by a particular type of structure. The embryo of the chick may agree with adult lower animals in its grade of development, but the first part that is differentiated in the chick is the chorda dorsalis, from which it is plain that the chick does not correspond to any invertebrate animal: "the type of every animal appears to be fixed in the embryo from the very first, and to regulate the whole course of development." Nor does the chick pass through the forms of the lower vertebrate animals. The embryos of all vertebrate animals are clearly vertebrates themselves, for they exhibit the chorda dorsalis and other features peculiar to their type. But no adult vertebrate animal is so little differentiated as vertebrate embryos. There is then no recapitulation of lower forms even within the limits of a single type.[13]

Von Baer formulated what he called the true law of development on the basis of further observations on vertebrates. The embryos of mammals, birds, and reptiles are so alike at their earliest stages, he said, that it is next to impossible to distinguish among them. Only later, with the development of their extremities, does it become clear to which class they belong. The first characters that appear in development mark only the greater divisions of the Vertebrata; those that appear later mark the smaller divisions. The more special form is developed from the more general. If this is true, is it not likely, von Baer asked, that at a very early period all animals have a common form? His answer was that the simple vesicle – the form of the germ – is that common fundamental form. At this, the earliest stage of development, the embryo is simply an animal. Subsequently the characters appear which show that it belongs to one of the four principal types, and it assumes the form of, for instance, the vertebrate archetype. Still later it acquires the characters of its class, family, and finally species. "The development of the embryo with regard to the type of organization, is as if it passed through the animal after the manner of the so-called *méthode analytique* of the French systematists, continually separating itself from its allies, and at the same time passing from a lower to a higher stage of development." Von Baer's four-part law of development is his general statement of this process of specialization: (1) The more general characters of a large group of animals appear earlier in their embryos than the more special characters; (2) from the most general forms the less general are developed, and so on, until finally the most special arises; (3) every embryo of a given animal form, instead of passing through the other forms, rather becomes separated from them; (4) fundamentally, therefore, the embryo of a higher form never resembles any other adult form, but only its embryo.[14]

Von Baer pointed out that in this process there will necessarily be resemblances between the embryos of higher and the adults of lower forms. These are correspondences in grade of development between vertebrate embryos and the adult forms of some lower invertebrates. And there will be still more fundamental resemblances between the embryos of higher vertebrates and the adults of lower vertebrates because the adult fish and the mammalian embryo, for instance, share both a common type of structure and a low grade of development. This is simply a consequence of the fact that within each type there are lower and higher grades of development – the adult mammal is further differentiated from the vertebrate archetype (the lowest grade of vertebrate development) than is the adult fish. But there is no recapitulation. The embryonic mammal, though it may resemble a fish, never assumes the form of a fish, for the adult fish has its own special characters which the embryonic mammal never acquires.[15] The nature of development is divergence from common forms – the simplest vesicle, then the vertebrate archetype, then the common forms of birds, reptiles, and mammals, and so forth – not passage along a linear series.

E. S. Russell, who was almost as fond of von Baer as he was of Cuvier, praised von Baer by comparing his functional approach to Cuvier's: "Von Baer, like Cuvier, never forgot that he was working with living things; he was saturated, like Cuvier, with the sense of their functional adaptedness."[16] This sense of the adaptedness of organisms is apparent in von Baer's insistence that the mammalian embryo is never a fish: The fish is not merely "an imperfectly developed Vertebrate," for it, as well as the mammal, has its own special characters which fit it for its mode of existence. But on the whole von Baer's approach more closely resembles Geoffroy's than Cuvier's. The two relations of organisms identified by von Baer – grade of development and type of organization – are both antithetical to Cuvier's strictly teleological conception of organic nature. Cuvier believed that perfection of organization has relation only to the ends to be achieved, and since the functional requirements of no two animals are the same, each animal must be considered apart. Assigning different degrees of perfection on the basis of some one organ or system of organs only results in the neglect of other characters. In Cuvier's mind, the idea of degrees of perfection was bound up with the idea of the animal series, to which he was unalterably opposed. So, while he recognized that in terms of overall complexity some organisms were higher than others, he said that it was impossible to classify animals in such a way as to indicate the perfection of forms relative to one another.[17] For von Baer, on the other hand, "grade of development" was a fundamental component of his view of the animal kingdom. His awareness of the adaptedness of each animal in no way interfered with his ability to recognize different degrees of perfection within each group.

Von Baer's concept of the archetype still more decisively distinguishes his views from Cuvier's and reveals the similarity between his outlook and Geoffroy's. For Cuvier there was no unity of type, and consequently no archetype, even within the confines of a single *embranchement*. If all vertebrates are similar, it is not because all are modifications of a fundamental vertebrate type of structure, but rather because all must satisfy similar conditions of existence. "It is necessary," he said, "to consider each being, each group of beings, in itself, and in the role which it plays by its characteristics and its organization, and not to make abstraction of any of its relations, or any of the bonds which connect it to other beings, either the nearest or the most distant."[18] But such abstractions – the archetypes – are the essence of von Baer's theory of animal organization. The contrast could not be clearer between

Cuvier's view and his: "every portion [of an animal]," von Baer said, "is to be understood only by its relation to the type and by its development out of it."[19]

Von Baer's outlook is in many ways typical of the generation of zoologists that began writing in the 1820s and 1830s. His commitment to the concept of archetypes places him squarely in the morphological tradition of Geoffroy and the German transcendental anatomists. But his views on the mutual relations of animals were intermediate between the extremes of Geoffroy and Cuvier. He accepted neither the idea of the unity of the whole animal kingdom nor Cuvier's denial of all unity. Von Baer perceived nature as a combination of diversity and underlying unity, with stringent limits placed on the extent to which such unity could be traced. In this perception, which received its clearest expression in his description of development, von Baer laid down the main lines of the branching conception of the organic world.[20]

Embryology and classification: Milne Edwards

Von Baer suggested that because during development characters appear in their order of generality in the animal kingdom – that is, those of the type, then the class, the order, and so on – embryology provided a key to natural classification. But he indicated only briefly how to apply his view of development to the problem of determining affinities, and when he did so, his main concern was to demonstrate its incompatibility with any arrangement of organisms in a series.[21] Others carried the argument much further. Martin Barry's articles "On the Unity of Structure in the Animal Kingdom" have already been mentioned in connection with the post-Cuvier rejection of the strictly teleological approach in comparative anatomy. In them Barry attempted to provide a synthesis of the most recent research in embryology. As part of his discussion he pointed out the great importance of the study of development for natural history generally, and he concluded with an explanation of its utility in classification.

Barry's heavy reliance on von Baer is apparent throughout his articles.[22] "A law," he said, "not less vast in its importance, than it seems to be general in its application, may be supposed to direct structure in the animal kingdom": "This law requires that a heterogeneous or special structure, shall arise only out of one more homogeneous or general; and this by gradual change. The importance of this law seems to have been insisted on chiefly by Von Bär, who arrived at it by long and attentive observation of development."[23] In all major points, Barry's description of development is the same as von Baer's. He followed von Baer in distinguishing between the degree of development and the type of organization. And like von Baer, he denied that the embryos of higher animals repeat the forms of the lower animals. Such repetition required a single scale of structure differing in degree alone, but no such scale exists. "Strictly speaking, therefore, no animal absolutely *repeats* in its development, the structure of any part of any other animal." The human embryo is never anything but a human embryo; the tadpole is never a fish, despite resemblances in form and habit.[24]

In introducing the subject of classification, Barry said that it seemed as if, with the "original design to create organized beings, there had arisen a scheme" of division and subdivision down to the level of species. A "perfect" arrangement would be that very scheme, but such an arrangement is not likely to be discovered by the use of adult characters alone. It is a corollary of von Baer's view of development, he said, that it is only at an early stage that "structure presents itself alone." In adult animals "function tends to embarass"; that is, special adaptive structures disguise the underlying similarities of form. For this reason embryology is the only sure basis for classification, the best hope for discovering the perfect arrangement: "The tree of animal development" reveals the "*grown tree* of *animal* structure."[25]

A few years later, in an article devoted exclusively to embryology and classification, Milne Edwards presented a much fuller justification for basing classification on embryological characters, and he explained at some length his reasons for believing that the diverging paths of development in the animal kingdom suggested by von Baer's theory correspond precisely to the branching series of organisms in a natural classification. Like Barry, Milne Edwards derived the image of a branching tree from the theory of diverging development. In an early paper, "Observations sur les Changemens de Forme que Divers Crustacés Eprouvent dans le Jeune Age" (1835), he had set forth a theory of development essentially the same as von Baer's, though he indicated subsequently that he did not at that time know of von Baer's work. In it he said that the changes which crustaceans undergo in the course of development tend to remove them ever further from the type common to the group to which they belong. He defined the common type as "an ideal and abstract form which represents all that which the different organisms have in common and the mean of the differences which distinguish them from one another." Development proceeds from the more general to the more special form, so that the particular characters of a species or genus either do not exist at all among the young or at least are much less pronounced than among the adults.[26] The idea with which he opened his memoir of 1835, that embryology could be of value in perfecting the natural method of classification, was expanded in his "Considérations sur quelques Principes Relatifs à la Classification Naturelle des Animaux" of 1844. A natural classification, he said, should arrange animals according to their degree of zoological relationship; that is, distribute them so that the most similar are closest together and the distance between animals is a measure of their differences. And it should subdivide the groups according to the differences in the constitutions of their members. Often it is easy to recognize affinities, but sometimes they become vague, and sometimes they are obscured by striking modifications which catch the eye even though they have no great significance. This causes difficulties because of the method naturalists commonly employ, namely, considering only adult organisms and neglecting the embryonic stages through which they pass. It is the process of development, he said, that reveals natural affinities with the greatest certainty.[27]

Milne Edwards referred to his early memoir on crustacean development for the demonstration that development consists in departure from the common type. Characters common to the family are acquired before those of the genus, and those of the genus before those of the species, a result that accords perfectly, he said, with von Baer's principles. From this it is easy to prove that "the general tendency of nature is to make the duration of these primordial resemblances of organisms in the course of development correspond with the various degrees of parentage or zoological affinity that the animal species preserve among themselves when they have arrived at the conclusion of their development." The germs and embryos of different animals may never be identical, but they are similar, and the similarity is ever greater as

Henri Milne Edwards. (National Library of Medicine.)

one approaches the initial stages of development. It is from this correspondence between degrees of relationship and the period of embryonic divergence that the utility of embryology in classification arises. The points in which different animals resemble each other represent an ever smaller proportion of their zoological characters as they approach maturity. Consequently , it is much easier to discover natural affinities by means of the history of development than by the study of mature organisms, in which the characters indicative of affinity make up only a small part of the total. It is in the embryo, Milne Edwards said, that one must seek the characters essential to the great zoological groups, while it is in the fully developed animal, generally the male adult, that one finds the characters of the species.[28]

Milne Edwards then discussed the kind of natural groups that embryology discloses. Like von Baer, he noted the relationship between the theory of recapitulation and the idea of a single scale of organic perfection, both of which he rejected. But he said that in denying the existence of an animal series it is not necessary to follow Cuvier's example and reject also the idea of a natural classification that recognizes different degrees of organic perfection. Embryology shows that there is no single series because each species, in diverging from the common embryonic form, follows its own path of development. Rather than a linear series, the metamorphoses of embryos of the animal kingdom form a multitude of series united at their base and branching off at different levels. These series formed by the stages through which the embryos pass are strictly parallel to the various natural series of adult animals. In development each species separates itself more or less from its neighbors, but this divergence occurs only in the very last stages of development and results in only slight differences, such as those which distinguish species and genera. Embryos that develop along a particular path will therefore resemble successively the adults that diverge to their special forms along the way. And these adults must then form natural series that correspond closely to the series of embryonic stages of those animals whose development proceeds furthest along that same path. The result is a branching arrangement. The series of adults, like the series of embryonic stages through which the various animals pass, represent a sort of tree, Milne Edwards said, "a tree which in rising from the ground separates into several stems each of which then divides into secondary main branches and terminates in innumerable little branches; but like the leaves with which a tree is covered, the species of animals thus produced can never, without flagrant violation of their natural relationships, be ranged in a single line." In this tree there is both branching and an increase in perfection. Development, though it is sometimes retrograde, generally results in an adult more highly organized than its embryo. The development of the embryo, that is, usually consists in an increasing division of physiological labor, which was Milne Edwards's criterion of perfection. Similarly in the corresponding series of adults, those that have developed furthest before diverging to their special forms are characterized by a greater division of physiological labor than those that diverged earlier. Animals but little developed from the common forms of their types and subtypes are located near the bottoms of the main stems and branches of the tree, while the more perfect forms are at the top.[29]

It is easy from this to see why Darwin thought highly of Milne Edwards's essay.[30] The branching arrangement Milne Edwards proposed agreed far better with Darwin's views than any other scheme of classification then current. But though there would seem to be a ready harmony between Darwin's theory and the ideas on classification proposed by Milne Edwards, I will show in Chapter 7 that Milne Edwards's essay in fact posed questions for Darwin that cost him a very considerable effort to resolve.

Richard Owen, the vertebrate archetype, and the fossil record

Milne Edwards, like von Baer, occupied an intermediate position between Geoffroy and Cuvier. He criticized those who used embryology to uphold Geoffroy's idea of the unity of type of the whole animal kingdom.[31] But he also found fault with Cuvier for denying the possibility of any classification based on varying degrees of perfection. For him, as for von Baer, grade of development was an important factor to be taken into account in determining relations of organisms. In his general approach to the study of animal organization, he sided with Geoffroy, employing in all his work the concept of unity of type within the limits of each *embranchement.* Richard Owen also chose this middle ground. He, more explicitly than von Baer or Milne Edwards, urged that comparative anatomy should proceed by taking the best of both Cuvier and Geoffroy.[32] Owen's eclecticism resulted in one of his most useful contributions to the study of comparative anatomy, the first clear statement of the distinction between analogy and homology. In his lectures at the Royal College of Surgeons in 1841 Owen said that every part of the skeleton must be studied in "two distinct but by no means incompatible points of view – to ascertain first to what it may be analogous & secondly with what it is homologous – i.e. to study the skeleton teleologically & morphologically – or with the view of ascertaining what each part *is* and to what end it is *adapted.*" His published definitions of "analogue" and "homologue" correspond to this distinction between the two methods of study.[33] Owen presented his compromise as a halfway point between the pure teleology of Charles Bell and the pure morphology of Geoffroy, but in fact the distinction has meaning only for the morphologist. The strict teleologist cannot admit it. For Cuvier (or Bell), all cases in which "the same organ" appears in different animals are cases in which they have the "same function," for similarity of ends, or conditions of existence, is the only admitted cause of similarity of form. For the morphologist, on the other hand, who believes the "same organ" may be modified to perform different functions in different animals, it is essential to know when two organs are "the same" and when they have merely an adaptive similarity.

Owen's inclination toward the morphological perspective of Geoffroy was stimulated by Barry's articles, which convinced him that embryology held the key to the law of unity of type.[34] He concluded his published lectures of 1843 with a von Baerian account of comparative development of animals, which he offered as an attempt to give "a more exact enunciation" of the "extent to which the law of 'Unity of Organisation' may . . . justly, and without perversion of terms, be predicated of animal structures." Following Barry, Owen said that the degree of resemblance between different animals is in inverse ratio to their approximation to maturity. All animals resemble each other at the earliest period of development, when they all correspond, though not in every detail, to the "polygastric animalcule": "The potential germ of the Mammal can be compared, in form and vital actions with the Monad alone, and, at this period, unity of organisation may be predicated of the two extremes of the Animal Kingdom." The infusorial monad is the fundamental or primary form of the animal kingdom. From that point development diverges to forms

Richard Owen.

common to each of the main types of organization. These forms
are secondary in relation to the animal kingdom as a whole, but
primary in relation to their *embranchements*. Radiates, Owen said,
pass from the monad stage to that of the polyp; articulates to that
of the worm; molluscs to that of the tunicate (which he classed

among molluscs); and vertebrates to the form of the finless cartilaginous fish. Within each type as well, radiation from common forms is the rule. Molluscs, for instance, after attaining the common form of their type, "either remain to work out the perfections of that stage, or diverge to achieve the development of shells, of a head, of a ventral foot, or of cephalic arms." Unity of organization prevails at early periods, only to be "lost in the diversity of the special forms as development proceeds":

> Thus every animal in the course of its development typifies or represents some of the permanent forms of animals inferior to itself; but it does not represent all the inferior forms, nor acquire the organization of any of the forms which it transitorily represents. Had the animal kingdom formed, as was once supposed, a single and continuous chain of being progressively ascending from the Monad to Man, unity of organisation might then have been maintained by the disciples of the Geoffroyan school.[35]

Embryology was useful to Owen chiefly for showing the limits of the law of unity of type.[36] But it also provided him with a developmental model which exerted a significant influence on his most important theoretical work – the delineation of the vertebrate archetype and the application of the concept of archetypes to the interpretation of the fossil record. Owen argued that the concept of an archetype was the only acceptable explanation of homologies once it was recognized that the "teleological hypothesis" could not satisfactorily account for them. We are forced to recognize, he said, an ideal pattern "on which it has pleased the divine Architect to build up certain of his diversified living works." Otherwise one would have to consider homologies matters of chance, a suggestion no "reasonable mind" would entertain.[37] In his *On the Archetype and Homologies of the Vertebrate Skeleton* Owen said that to discover the archetype it was necessary to compare all the members of the vertebrate type, in order to find each part in its least modified state.[38] The form he described was in effect the lowest common denominator of the vertebrates, an unspecialized form with the potential to develop into any vertebrate animal. It was composed of an indefinite number of basically similar segments, each of which Owen called a vertebra:

> The natural arrangement of the parts of the endoskeleton is in a series of segments succeeding each other in the axis of the body . . . For each of these primary segments I retain the term "vertebra" . . .

I define a vertebra, as *one of those segments of the endo-skeleton which constitute the axis of the body, and the protecting canals of the nervous and vascular trunks:* such a segment may also support *diverging appendages.*

The head Owen considered to be composed of four greatly modified vertebrae. The fins, paddles, wings, legs, and arms of vertebrate animals are the diverging appendages of two vertebrae, four limbs being the usual number in vertebrates on this planet, although elsewhere there might be more.[39]

Owen's archetype, like von Baer's, was the lowest grade of vertebrate development. It was characterized by the "vegetative" or "irrelative" repetition of similar parts, without their being combined or modified to perform special functions (in von Baer's terms, there is little morphological differentiation). This vegetative repetition Owen imagined to be produced by the operation of a "polarizing force," the same as that which causes the growth of crystals. The polarizing force is one of two fundamental forces at work in the organic world. Each animal owes its special form to the second, a teleological force – "the adaptive or special organizing force" – which modifies the similar segments of the archetype in various ways according to the animal's needs. In animals but little removed from the archetype, the "vegetative-repetition-force" is predominant. The more the archetypal form is obscured by adaptive modifications produced by the "adaptive force," the higher the grade of development.[40]

An archetype composed of a series of ideally identical vertebral segments could never actually exist, for want of any adaptive modifications. But the archetype Owen figured in his works is clearly molded to some degree by the "adaptive force." The neural arches of the four anterior vertebrae are enlarged into the beginnings of a cranium and are perforated for the nerves of the sense organs. The posterior vertebrae diminish in size to form a tail. And there are both a mouth opening and a vent.[41] Ideally the archetype was, in Owen's view, the "stem" from which the various series of vertebrate animals diverged.[42] But it appears that he had in mind also a real, vaguely amphioxus-like animal from which the vertebrates might actually have developed through time. Whether or not this was his opinion when he prepared his "Report" on the archetype for the British Association in 1846, five years later Owen had constructed on the basis of his morphological ideas a general theory of paleontological succession.

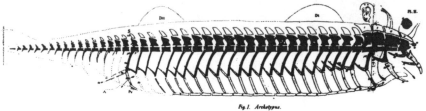

Fig. I. *Archetypus.*

Owen's representation of the vertebrate archetype. (*On the Archetype and Homologies of the Vertebrate Skeleton*, 1848, plate 2.)

Owen's career as a paleontologist began with his work on the fossil mammalia collected by Darwin in South America. From an early period three sorts of fossil forms attracted his attention. One was those forms that seemed to combine in a single animal characters that later appeared in two or more separate groups. In his first paleontological paper he characterized the *Toxodon* as a rodent with affinities to the pachyderms (which then included all the nonruminant hoofed mammals) and the "herbivorous cetacea" (dugongs, manatees). In describing his reconstruction of *Mylodon robustus*, he ranked the giant sloth as an unguiculate which provided a transitional link to the other great mammalian series, the ungulates. His several contributions to the *Zoology of the Voyage of H.M.S. Beagle* conclude in the same vein. The *Macrauchenia*, for instance, seemed to link the pachyderms and ruminants: It is "in a remarkable degree a transitional form, and manifests characters which connect it both with the Tapir and the Llama."[43] In 1847 Owen substantially revised Cuvier's classification of the ungulates, replacing his ruminants and pachyderms with the subgroups Artiodactyla (even-toed) and Perissodactyla (odd-toed). On the basis of the connecting links formed by a number of extinct animals, including *Macrauchenia*, he placed in the former group not only the ruminants of Cuvier, but a large number of the pachyderms as well.[44]

Owen was not alone in focusing attention on such transitional forms. All paleontologists recognized them as being of particular interest. William Buckland said they supplied links in the chain of being. Louis Agassiz called them "prophetic types" and said they pointed toward parts of the plan of creation that were not realized till a later period. Even MacLeay, who was no paleontologist, suggested that some of his osculant types – such as that connecting birds and mammals – would probably be found only in a fossil state.[45] But these forms took on a new significance when interpreted in the light of the branching conception. Then they appeared as points from which two or more series of animals have diverged.

Other fossils seemed to be transitions in linear, chronological series, filling gaps, for instance, between the *Palaeotherium* of the Eocene and the modern horse. The fossils in this particular series were described by a number of paleontologists in the first half of the century. Owen's contribution was to point out how they formed an apparently developmental sequence, the teeth being modified in a determinate direction, the number of toes decreasing from three to one.

A third sort of fossil seemed to resemble embryonic stages in the development of living animals. These too were observed by many paleontologists, foremost among them Agassiz. The idea of a parallel between embryonic development and the geological history of animals occasionally was suggested by advocates of the theory of recapitulation in the early years of the century.[46] In the 1840s it was revived and made respectable by Agassiz, who with his assistant Carl Vogt found paleontological and embryological evidence to support it. They discovered an agreement, in the shape of the tail, between the embryos of modern fish and the adults of ancient orders of the class. Fish with heterocercal tails appear early in the fossil record, while fish with homocercal tails, such as modern bony fish, appear later. Following the same order of appearance, the embryos of homocercal fish are heterocercal before they acquire their adult form.[47] Later Agassiz added many similar instances drawn from other classes of vertebrates and invertebrates. Agassiz followed von Baer in the belief that there could be no transitions from one *embranchement* to another. But within a single *embranchement*[48] he said that the embryos of the higher animals pass through the forms of adults lower in the scale and that they also repeat the chronological sequence of extinct animals from the first appearance of the group in the fossil record. He postulated a threefold parallel between zoological affinities, geological succession, and embryonic development: "*The embryo of the fish during its development, the class of existing fishes in its numerous families, and the fish type in its planetary history traverse in all respects analogous phases, through which one always follows the same creative thought.*"[49]

Owen was as interested as Agassiz in the apparent parallel between embryonic development and the series of fossil forms. In his "Report on British Fossil Reptiles" of 1841 he pointed out that the reptiles presented analogies similar to those Agassiz had noted between ancient heterocercal fish and the embryos of existing homocercal fish. Two years later he suggested that this phenomenon was very common: "As we advance in our survey of the organisation and metamorphoses of animals, we shall meet with many examples, in which the embryonic forms and conditions of structure of existing species have, at former periods, been persistent and common, and represented by mature and procreative species, sometimes upon a gigantic scale."[50] But unlike Agassiz, Owen did not see this as recapitulation. The resemblances between embryos and fossils were precisely the same sort as the resemblances von Baer, Milne Edwards, and he himself recognized between lower adults and embryos of higher forms – as, for instance, the fish and embryonic mammal. Both fish and mammalian embryos are modifications of the vertebrate archetype, and both represent a low grade of development within the type, so there must inevitably be resemblances between them. But the embryonic mammal is never identical to an adult fish. It is just that its divergence from the archetype has not yet proceeded very far, has not yet caused it to be widely separated from the fish, which even as an adult is not greatly modified from the form of the archetype.[51] For Owen, the analogy between embryos and fossils indicated that the same law of development could be discerned in embryology and paleontology – namely, von Baer's law of development from the more general to the more special form.

In the 1851 edition of his *Principles of Comparative Physiology*, William B. Carpenter, drawing chiefly on Owen's work, discussed what he called the "application of Von Bär's law of *Development from the General to the Special*, to the interpretation of the succession of Organic forms presented in Geological time." Later the same year, and again in 1853, Owen presented his *"Law of Progression from the General to the Particular."*[52] Both Owen and Carpenter saw all three of the fossil types just discussed – those that combined the characters of succeeding groups and those that formed part of a linear series as well as those that resembled embryos of existing forms – as manifestations of this one law of development. All three represented a departure from the archetype of the group. In explaining what this meant in paleontological terms, Owen drew examples from invertebrates, fish, and reptiles, but the heart of his argument was the succession of hoofed mammals. The archetypes of the four great groups of animals, he said, are characterized by vegetative repetition, such as the repetition of similar vertebral segments. In such a highly developed group as the mammals, special adaptations have greatly obscured the form of the archetype, but even here some vestiges of vegetative repetition remain. Thus, the archetype of the mammals, or the "general mammalian type," has five toes on each foot and forty-four teeth. All existing adult hoofed mammals have special modifications which cause them to depart more or less from this common type, but these special modifications are acquired during the development of the embryo from a more general form. Like these embryos of the existing species, extinct species generally departed less widely from the type. The *Anoplotherium*, for instance, had the typical forty-four teeth, and they were less specialized for particular functions than the teeth of existing ruminants. The earliest known rhinoceros had fully developed incisors in both jaws and four toes on each foot, but modern rhinoceroses have lost the incisor and have only three toes. Owen illustrated the same process of specialization in the European series of horselike mammals. Compared to any known species of *Equus*, he said, the *Palaeotherium* of the Eocene departs less from the common type in having three functional toes and in retaining the first premolar in both jaws. The Miocene *Palaeotherium* departs further from the type, its outer and inner toes being greatly reduced in size. Another Miocene species, *Hipparion*, has these toes even further reduced. Only in the Pliocene are species of *Equus* found, in which the second and fourth toes are mere rudiments. Mammalia, Owen said, have departed ever further from the common type during the Tertiary period:

> The above cited and other analogous facts indicate that in the successive development of the mammalia, as we trace them from the earliest tertiary period to the present time, there has been a gradual exchange of a more general for a more special type. The modifications which constitute the departure from the general type adapt the creature to special actions, and usually confer upon it special powers.

The horse is the swifter by reason of the reduction of its toes to the condition of the single-hoofed foot; and the antelope, in like manner, gains in speed by the coalescence of two of its originally distinct bones into one firm cannon-bone.[53]

While in the linear succession of rhinoceroses and horselike mammals there appeared to be progressive specialization; the various mammalian series seemed to diverge from the common type: The Miocene rhinoceros and its contemporary the *Tapirotherium* more closely resembled each other than do the modern rhinoceros and tapir. Similarly, such transitional forms as *Toxodon* and *Mylodon robustus* linked series of animals whose modern representatives are widely separated.[54]

Owen's paleontology is a far cry from that of Cuvier or of Cuvier's progressionist followers, such as Buckland. Cuvier established the fact of a succession of organic forms in the history of the earth, and Buckland said that successive geological periods were characterized by more perfect forms, perfection having some vague reference to organic complexity and to the present creation and man. In place of their vision of successive stages, Owen substituted a vision of organic development analogous to divergent embryological development. There is an obvious sense in which Owen's conception of the history of life is protoevolutionary. It might be said to be ready-made for reinterpretation in harmony with the doctrine of descent.[55]

At a later period both Owen and Huxley assessed the progress of paleontology, and though their remarks were intended to glorify their own efforts and vindicate their past judgments, and though they reflect their different attitudes toward Darwin, they nevertheless offer, if read cautiously, a useful comment on the shift that occurred in the generation after Cuvier. Huxley in 1880 wrote, "If the doctrine of evolution had not existed, palaeontologists must have invented it, so irresistibly is it forced upon the mind by the study of the remains of the Tertiary mammalia which have been brought to light since 1859."[56] Much indeed was discovered after 1859.[57] But Huxley's reason for singling out that year had to do more with the history of his personal views on evolution than with the history of paleontology. Until 1859 Huxley argued vehemently that the evidence then available did not support any doctrine of transmutation or of progressive development.[58] His own conclusions of 1876 regarding some of

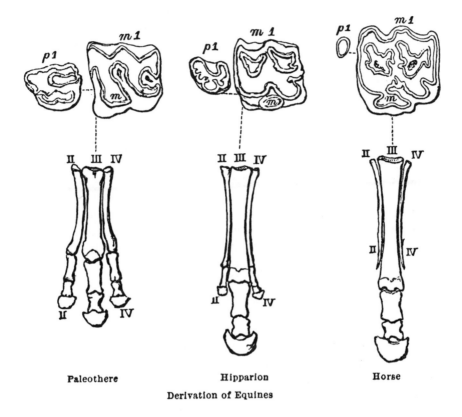

Paleothere   Hipparion   Horse

Derivation of Equines

Owen's illustrations of specialization in the ancestry of the horse, originally used in his paleontological lectures at the Royal School of Mines, 1857. (*On the Anatomy of Vertebrates*, 3 [1868]: 825.)

the tertiary mammalia suggest otherwise. In his *American Addresses* Huxley claimed that the European series of horselike mammals provided as clear proof of evolution as any "merely inductive conclusions" could ever receive. Although Huxley did not of course acknowledge the fact, this "Demonstrative Evidence of Evolution" was none other than the same fossil series that Owen had discussed in 1851.[59]

Owen, in his retrospective of 1868 on his own contributions to the establishment of the doctrine of descent, called attention to the way in which his work and that of others had satisfied the conditions laid down by Cuvier: "To the hypothesis that existing are modifications of extinct species Cuvier replied . . . 'You ought . . . to be able to show, e.g., the intermediate forms between the Palaeotherium and existing hoofed quadrupeds.'" The progress of paleontology since 1830, Owen said, had supplied many "missing links," which, when rightly interpreted, as by himself, favored the "idea of organisation by secondary law." The characters of the *Anoplotherium,* which seemed so anomalous in the time of Cuvier, "lost much of their import as evidence of insulated form, or special creation, when they came to be viewed by the light of the law of the 'more generalised character of extinct species.'" "Thus amply and satisfactorily," he said, "has been fulfilled Cuvier's requisition of 1821."[60]

Here Owen and Huxley, despite themselves, are made to agree on the significance of the reinterpretation of the fossil record that occurred in the two or three decades before 1859. Both are wrong in one important respect. Neither the *Palaeotherium* – horse series nor the law of development from general to special was able to persuade either of them to proclaim publicly, nor Huxley to profess privately, a belief in descent. Darwin did that. But there may be more than a grain of truth in Huxley's unintentional suggestion that the course that paleontology was on by the 1840s would have established the doctrine of descent even without the *Origin.*

Mid-century morphology and British natural theology

So far I have been discussing the branching conception simply as a new development in natural history. But its impact was felt well beyond the narrow circle of professional biologists, largely because of the special place organic nature had always held in natural theology.[61] I noted in Chapter 1 that Owen and others who rejected the strict teleological interpretation of organisms were concerned to argue that this move did not endanger natural religion. The necessity of their making this claim derived from the related facts that the teleological method in comparative anatomy was a bulwark against transmutationism and that the morphological approach was widely perceived as undermining the best proof of design. If an organism is constructed as it is in order to conform to a plan, rather than to serve a particular function, then is it not made without any purpose – which is to say wastefully and unreasonably? And if every part must conform to a plan of structure, its perfection must be limited, because the ways in which it is fitted to its function are circumscribed by the law of adherence to type.

Owen's *On the Nature of Limbs*, originally a Friday evening lecture at the Royal Institution, had as one of its major aims the refutation of such objections:

> Those physiologists who admit no other principle to have governed the construction of living beings than the exclusive and absolute adaptation of every part to its function, are apt to object to such remarks as have been offered . . . that 'nothing is made in vain;' and they deem that adage a sufficient refutation of the idea that . . . apparently superfluous bones and joints should exist in their particular order and collocation in subordination to another principle; conceiving, quite gratuitously in my opinion, the idea of conformity to type to be opposed to the idea of design.[62]

Owen's formulation of some of his principal biological doctrines illustrates the way in which personal belief, social and professional requirements,[63] and more narrowly biological concerns shaped his science. His distinctions between "adaptive" and "essential" characters and between analogy and homology; his hypothesis that form is governed by two forces, a teleological force responsible for adaptation and a "polarizing" force responsible for the repetition of similar basic elements of structure; and his instruction to his students to study the organism both teleologically and morphologically are all parts of his attempt to forge a compromise that would make it possible to take advantage of the insights of Geoffroy and the German morphologists without abandoning entirely either Cuvier's principles or the argument that adaptation to purpose shows design in the creation. Owen continued to hold that adaptation is evidence of an intelligent creator, insisting that the fact that adaptation is produced by "natural laws or secondary causes" merely increases our admiration for the creator's skill in adapting the same elements of structure to so many different functions.[64] But in *On the Nature of Limbs* his main argument for the existence of God was an argument from plan, not purpose. His "discovery" of the archetype, he said, implied that its creator knew beforehand all the modifications of it that have successively appeared on earth.[65]

As a statement by Britain's foremost biologist on a subject of major importance in the mid-century debates over Chambers's *Vestiges* and over the most appropriate strategy for natural theology,[66] *On the Nature of Limbs* was frequently cited in the decade after its publication, and its claim that unity of type and design are not incompatible was widely accepted. Baden-Powell, for instance, drew on Owen to support his argument that the principle of the uniformity of nature must be applied even in the study of the history of life ("a real break in the connection and continuity of physical causes cannot exist in the nature of things"); and he cited Owen to show that natural theology would do well to rest its argument for design less on adaptation to purpose and more on the evidence of order and plan in nature.[67]

The case of William Whewell is still more revealing of the repercussions of the post-Cuvierian synthesis. In 1838 Whewell was the principal target of Carpenter's attack on teleological method. But in his anonymous *Of the Plurality of Worlds* of 1853, Whewell, saying he had "learnt much from Mr. Owen," reversed his earlier position on the sufficiency of the teleological method in biology and acknowledged the importance of the law of unity of type. At the same time, he followed Owen in recasting and reducing the importance of the argument from purpose as a proof of design. John Brooke has suggested that a main point of Whewell's argument against the probability that other worlds are inhabited was to deny that there is any *universal* law of development such as postulated by Chambers in *Vestiges*.[68] Since natural theology from the seventeenth century on had assumed that other worlds are inhabited, this raised the difficult question of what purpose other celestial bodies serve if they are not seats of intelligent life. Whewell answered that large portions of nature must be omitted from the argument from design, and he justified his response by reference to Owen's work:

Our naturalists and comparative anatomists, it would seem, cannot point out any definite end, which is answered by making so many classes of animals on this one vertebrate plan. And since they cannot do this, and since we cannot tell why animals are so made, we must be content to say that we do not know; and therefore, to leave this feature in the structure of animals out of our argument for design. Hence we do not say that the making of beasts, birds, and fishes, on the same vertebrate plan, proves design in the Creator, in any way in which we can understand design. That plan is not of itself a proof of design; it is something in addition to the proofs of design; a general law of the animal creation, established, it may be, for some other reason. But this common plan being given, we can discern and admire, in every kind of animal, the manner in which the common plan is adapted to the particular purposes which the animal's kind of life involves.

While Whewell was not willing to accept Owen's argument that the archetype proves design, he admitted that design was not incompatible with the law of unity of type. There are general laws in nature whose operation is wider than any particular purpose, he said, and yet there is also special adaptation to purpose.[69]

Whewell did not, like Baden-Powell, declare that the transmutation hypothesis is a valuable philosophical conjecture. But he appears to have recognized that in accepting Owen's verdict on the inadequacy of teleological explanations of structure he too had adopted a natural theological stance that would accommodate transmutation as one of nature's general laws.[70] There were no doubt many considerations that weighed with Whewell as he made this shift. From the conspicuous place Owen's arguments occupy in *Plurality of Worlds*, however, it appears that one of the most important was the recent progress of biology. Whewell shared what seems to have been a growing awareness that the new facts and ideas that were prominent in natural history had eroded many of the best scientific arguments against transmutationism.

Conclusion

With its extension to paleontology around 1850, the branching conception informed work in all the most important disciplines in natural history, except geographical distribution, for which it had no very direct significance. The natural history of the mid-nineteenth century was in consequence very different from that of Cuvier or the British naturalists of the period 1800–30. From the end of the eighteenth century, paleontology and geology contributed to a historical conception of the earth and its inhabitants; but the teleological assumption of a fixed relationship between organic form and inorganic conditions led Cuvier, Buckland, and others who shared their views to see progress as a sequence of discrete and in themselves static stages. Historians of biology, and particularly of paleontology, have tended to assume that this attitude persisted until 1859.[71] It would be most curious if this had been the case, for the first half of the nineteenth century saw developmental and historical modes of understanding and explanation invade most fields of thought; and their importance in German embryology from the time of C. F. Wolff has long been acknowledged.[72] The establishment of the branching conception marks the introduction of this same developmentalism into the several disciplines, centered around morphology, that formed the core of nineteenth-century natural history. That it took longer to make its appearance in the biology of France and Britain than in that of Germany is entirely consonant with what has been observed of developmentalism in general. If additional reasons are required to account for this difference in timing in biology, they most probably have to do with the character and strength of the reaction in France and Britain and with the prevalence in Britain of a mechanistic version of the design argument – in other words, with those conditions which, in conjunction with the impressiveness of Cuvier's work, gave strength to the program of teleological explanation.

Although it is necessary to recognize the connection of the branching conception with the rise of developmentalism and with the broad range of historical circumstances that produced it, it is important also to observe that the subjects of investigation, the data with which naturalists worked, and debates within biology contributed to the formation of and gave a particular shape to developmental conceptions in natural history. Embryology, in which development is most easily perceived, seemed to imply that the separate individuals and groups that the morphologists, taxonomists, and paleontologists dealt with might best be considered as stages of a developmental process (conceived usually in merely ideal terms). It is unlikely, however, that embryology alone or in concert with the general movement toward developmental conceptions of nature and society would have led naturalists to see development as a process of divergence. The idea of branching owed much to the increased perception of organic diversity that accompanied the birth of comparative anatomy.[73] It was this that led even Lamarck, who was committed to the animal series, to picture nature as branching. And it was this that served to justify, if not to motivate, Cuvier's criticism of the chain of being and of the doctrines of all-embracing unity held by Geoffroy, Oken, and others. Cuvier's teleological method, with its corollary that every organism is a functionally adapted whole, emphasized still further the diversity of nature. The result was that the essentially linear conception of development that prevailed in Germany before the work of von Baer was to a considerable extent replaced by the idea of development diverging from a common starting point. It is this aspect of the developmental conception as it was worked out in natural history before 1859 that is of particular importance in understanding Darwin's efforts in Part II of his species work.

As development in general was conceived of differently by different thinkers or schools – real or ideal, continuous or discontinuous, the product simply of law or the result of supernatural direction and activity – so too was organic development.[74] There were many theories of ideal development, many theories of transmutation, and still other theories of descent which denied gradual or transmutative change. And there were also numerous generalizations about the developmental patterns apparent in nature which avoided entirely the question of their cause. Von Baer considered his law of animal development to be but an aspect of a more general law: The "*one* fundamental thought which runs through all forms and grades of animal development, and regulates all their peculiar relations . . . is the same thought which collected the masses scattered through space into spheres, and united them into systems of suns; it is that which called forth into living forms the dust weathered from the surface of the metallic planet." For him, what was developing in all of nature was the world spirit; the purpose of development was the "progressive conquest of spirit over matter."[75] Owen and Milne Edwards tended toward the view of a plan of creation, with laws and design imposed by God.[76] These differences are worthy of serious investigation (they may help explain, for instance, the fact that Owen was more sympathetic to the doctrine of descent than was von Baer).[77] But for the purpose of understanding the development of Darwin's theory, it is the similarities that are most pertinent. Whether he was reading Owen, Milne Edwards, or Carpenter's or Johannes Müller's discussions of von Baer, Darwin found similar conceptions of archetypal forms; similar conceptions of divergence in embryology, taxonomy, and paleontology; and similar definitions of organic complexity and progress.

6

Darwin and the branching conception

The relationship of Darwin's theory to the branching conception of nature is at one level both obvious and well understood. As E. S. Russell put it, Darwin found that "the current morphology . . . could be taken over, lock, stock and barrel, to the evolutionary camp."[1] Darwin made the archetypes into ancestors, he argued that homologies and unity of type could be explained by descent, and he said that the ideal development from archetypal to specialized forms proposed by Owen and Carpenter had actually occurred during the history of the earth. Although this is obvious, it ought not to be taken for granted. To do so is to lose sight of the fact that Darwin's and his contemporaries' conceptions of evolutionary change were given a distinctive shape by their morphological perspective. Evolution does not have to be conceived as a development from generalized archetypal forms to more specialized types. This was not Lamarck's view. And it is not the view of the modern evolutionary synthesis: Ernst Mayr and others have been at pains to insist that ancestors are not necessarily, or even usually, archetypal forms.[2] But this was the view of virtually every evolutionist in the second half of the nineteenth century – a consequence of the conception of the structure and relations of organisms, both fossil and living, that came to dominate natural history in the generation after Lamarck and Cuvier. For Darwin, ancestors were in fact archetypes. Darwin emphatically rejected the notion of an "exemplar," a divine idea of a vertebrate animal that existed before the first actual vertebrate was created.[3] But the first vertebrate (which Darwin until around 1850 tended to think was *created*) was for him a generalized, archetypal creature. "I look at Owen's Archetypes," he said, "as more than ideal, as a real representation as far as the most consummate skill & loftiest generalization can represent the parent form of the Vertebrata."[4]

Unlike Lamarck, Darwin had more than his share of the "sense for morphology."[5] Late in his first transmutation notebook, when his serious reading in the literature of biology had just begun (two or three months after first examining the Cuvier – Geoffroy debate) he suggested that the type of each order might be "that form, which wandered least from ancestral form." In the second notebook he said that the abstract idea of a bird acquired by examining the whole class is a description of the "ancestor of all birds." And when he became acquainted with the vertebral theory of the skull, it led him to a vision of the ancestor of all vertebrates: "the head being six metamorphosed vertebra, the parents of all vertebrate animals must have been like some molluscous bisexual animal with a vertebra only & no head – !!"[6] From this it is apparent why Darwin found Owen's work so praiseworthy. It was in his eyes a rigorous and painstaking presentation of his own hypothetical view of the parent form of the whole vertebrate series. When Owen published his book on the vertebrate archetype, Darwin, while reading it, reiterated his idea of the ancestral vertebrate: "I cannot anymore doubt that some primordial race existed, long anterior to the lowest part of the Silurian System, which had a chain of vertebra & no skull, than I doubt the correctness of the nature of the skull & the beautiful explanation it affords of so many bones in it."[7] It is clear too why, despite Huxley's later criticism, Darwin continued to be attracted to the vertebral theory of the skull. The ideal metamorphosis morphologists traced from separate vertebrae to various forms of skull was for Darwin a description of the ways in which his ancient progenitor of the vertebrates had been modified in the course of evolution.[8]

Darwin did not merely borrow his morphology from others. He was in his own right a very able morphologist, as his volumes on cirripedes testify. The whole of his cirripede work is informed by the methods and goals of contemporary morphology. He was concerned with the tracing of homologies and the delineation of the "archetype Cirripede" and its relation to the more general archetype crustacean. And in part at least the research on barnacles was a search for the ancestral cirripede, which Darwin believed on morphological and geological grounds he had found in "Pollicipes, the old type-form of the whole order," the "stem of the genealogical tree," from which "all ordinary Cirripedes . . .

seem to radiate."[9] Had Darwin's career been cut off, as he often feared that it would be, before he published his theory of evolution, he would now be placed, on the basis of his investigations of cirripedes, among that large group of mid-nineteenth-century biologists who organized their work and thought around the idea of divergence from archetypal forms in each group of animals.

Besides this shared morphological perspective, there are other less obvious facets of the relationship between Darwin's work and the branching conception that are of some interest. Three of these in particular deserve comment. One is Darwin's belief that his analysis of organic structure – heredity plus adaptation – agreed "exactly" with the compromise Owen proposed between the views of Cuvier and Geoffroy. Another is that Darwin found it difficult to construct evolutionary explanations for some aspects of the branching conception, such as von Baer's laws, which on the surface appear to be readily compatible with the doctrine of descent. The third is the fact, too easily overlooked, that the new natural history appreciably strengthened the arguments that could be made for evolution.[10]

Morphology and teleology

In 1874 Asa Gray wrote in *Nature* that Darwin had done great service to natural science by "bringing back to it Teleology; so that, instead of Morphology *versus* Teleology, we shall have Morphology wedded to Teleology."[11] Darwin thanked him for the article in which the remark appeared and added, "What you say about Teleology pleases me especially."[12] Darwin was not merely being polite to his best American ally, who for years had been arguing for both natural selection and a refurbished argument from design. Since the mid-1840s Darwin had seen his theory as uniting the morphological and teleological approaches to the study of organic structure.[13] When in 1855 he found in Whewell's anonymous *Of the Plurality of Worlds* the statement that the best comparative anatomists – by which Whewell meant Owen specifically – say that researches on the structure of animals must be guided by the principle of "Unity of Composition, as well as the principle of Final Causes," his comment was "How exactly this agrees with descent &

selection for ~~final cause of~~ [*sic;* Darwin's cancel] advantageous structures."[14]

In the *Origin of Species* Darwin echoed Whewell's statement: "It is generally acknowledged that all organic beings have been formed on two great laws – Unity of Type, and the Conditions of Existence." His theory, he said, explains unity of type by descent. And it is compatible with Cuvier's principle because with minor exceptions every organic structure is an adaptation formed by natural selection either for the use of its present owner or for the use of some progenitor of its present owner. On his theory, he argued, the law of the conditions of existence is the higher law, because "it includes, through the inheritance of former adaptations, that of Unity of Type."[15] As Darwin well knew, this was not what Cuvier meant by his principle. According to Cuvier, the structure of every animal is wholly referrable to its own conditions of existence, not those of its ancestors. This Darwin had denied since 1837. For him the structure of every animal, ancient or modern, is chiefly due to heredity.[16] But in arguing for the compatibility of descent and the doctrine of the conditions of existence, Darwin was not being disingenuous. He believed, like Owen, that while he rejected Cuvier's principle as an all-sufficient explanation of form, he could still make use of it in a more restricted sense. When he said in the *Origin* that Cuvier's principle is "fully embraced" by natural selection, he was perhaps going too far. Elsewhere he expressed the same sentiment more cautiously, urging that his views agreed "sufficiently well" with Cuvier's.[17] It may be said, and it is probably partly true, that this claim was a ploy to undercut anticipated attacks by Cuvierians.[18] But it must be remembered that Darwin made essentially the same claim in reading-notes that were never intended for publication. It is true too that Darwin was always anxious to find that his theory could be made to agree with and explain the best contemporary biology, in this case Owen's morphology; and we may well conclude that this led him to gloss over some important differences between his views and those of both Cuvier and Owen. But Darwin's belief that his theory united morphology and teleology also reflects some deep-seated attitudes toward nature which he shared with most of his contemporaries – in particular, what he later admitted was a natural theological view of organic structure.

Darwin believed that all structures, with the possible exception of those produced by "correlation of growth" and the direct effect of conditions, originated as adaptations. In the first transmutation notebook, this was because there was a designed "law of adaptation." After 1838 Darwin supposed that natural selection was the law of adaptation, the agent by which adaptation is produced. In his notes on Owen's *Archetype and Homologies of the Vertebrate Skeleton* Darwin pictured selection as serving the same function as Owen's teleological principle. While Owen supposed that his "adaptive force" fashioned the elements of the archetype to the various needs of all the species built on that plan, Darwin said that selection does this. Owen observed that the separate bones of the skull, which have their origin in the archetype, are specially adapted in mammals to ease the passage of the head through the birth canal. A structure of independent origin, Darwin commented, "is usefully, nay indispensably worked in; it is found out & used by the selection principle."[19] In the *Descent of Man* Darwin explicitly identified as a theological assumption his long-held belief that every part of every organism is an adaptation. This suggests that the agreement he saw between his analysis of structure and Owen's was not merely superficial, a possibility that is strengthened by a remark Darwin made in his copy of Owen's *On the Nature of Limbs*. There he said that he agreed with Owen that unity of type and design are not incompatible.[20] Like Owen and many others in his generation, Darwin was concerned to explain fundamental organic relations without denying purpose; and from his often repeated statements that natural selection is a second cause and that he believed in design – of a sort – it is clear that in his view purpose derived from God. Natural selection was in this sense purposeful: it was a law instituted by God to produce adaptation.[21] And so, like Owen with his polarizing and adaptive forces, Darwin thought he could legitimately claim that heredity and selection combined the best of Geoffroy and Cuvier in a single theory.

What Gray called Darwin's reintroduction of teleology into biology was probably of little service in gaining the favor of the readers of the *Origin* ("I do not think any one else has ever noticed the point," Darwin said).[22] But in another way, touched on already, the similarity between Darwin's and Owen's outlooks may have been of considerable importance. Owen's compromise supported Darwin's major project – establishing the fact of evolution – by its insistence that the principle of the conditions of existence alone gave an inadequate account of structure, and by its concomitant sapping of the one version of the argument from design that was incompatible with the doctrine of descent. In Owen's "splendid sentences" on the inadequacy of explanations in terms of final causes Darwin found a welcome ally in his attack on the teleologists' antievolutionary interpretation of organic nature.[23]

The problem of von Baer's laws

If it is at first sight surprising to find Darwin arguing, against Huxley and Wallace, for "strict adaptation" and Cuvier's principle of the conditions of existence,[24] it is equally odd to discover that Darwin found it difficult to incorporate a branching model of embryonic development into his theory of branching evolution. Darwin saw the principal generalizations of morphology and paleontology as simple consequences of descent ("I sh[d] put it as self evident that descent explains unity of type"),[25] but this was not true of von Baer's laws. In his efforts to explain them we see how developments in contemporary biology could shape his perception of nature; and we see how Darwin's project of giving an evolutionary interpretation of current natural history could occasionally lead him into important lines of speculative activity.

Throughout the period covered by the transmutation notebooks (1837–9), Darwin conceived of embryonic development as a process of recapitulation, a "typical or shortened repetition" of the ancestry of the species.[26] Like all recapitulationists, he saw change as predominantly additive, the series of stages through which the embryo of a higher animal passes being produced by additions to the end of the developmental path followed by the embryo of the first and simplest ancestral form. This process of addition was an important component of Darwin's pre-Malthus theory of transmutation. He believed it was the repetition of stages in sexual generation that accounted for the "power of adaptation." It is necessary, he said, that something should be added each time in "that kind of generation, which typifies the whole course of *change* from simplest form," for the main purpose of that kind of generation is to allow the embryo to pass through the "whole series of forms to acquire differences."[27]

After Darwin formulated the theory of natural selection the role of the generative process in his evolutionary system was altered, but this had no effect on his conception of embryonic development. Some months after reading Malthus he entered in his fourth notebook the reflection that since all vertebrates can be traced to a germ, and since all are descended from one stock, there is reason to think that the birth of species and the birth of individuals are closely related: "By birth the successive modifications of structure being added to the germ, at a time (as even in childhood) when the organization is pliable, such modifications become as much fixed, as if added to old individuals during thousands of centuries, – each of us then is as old as the oldest animal, have passed through as many changes as has every species."[28] Man is at the end of the line of development that has been formed by the addition of stages to the simple condition of his earliest ancestor; and in his own development as an individual he passes through those stages.

It was not the theory of natural selection that led Darwin to give up his belief in recapitulation. Nor was it simply further reflection on his own conception of the upward branching of life (his first three notebooks are ample testimony to the compatibility of recapitulation with a branching model of descent). Darwin abandoned recapitulation when it began to lose favor among younger biologists whose opinions he respected. In 1837–9, when von Baer's alternative interpretation of development was just beginning to be widely known, Darwin had no trouble in finding authorities to bolster his belief in recapitulation. He knew that Geoffroy supported the theory and noted too that his disciple Serres, the principal French exponent of recapitulation, thought molluscs represented the fetuses of vertebrates. Owen told him this was nonsense, but not long afterwards Darwin read an earlier work by Owen in which he had said that the Acrita are analogous to the earliest embryonic stages of the higher classes of animals. Darwin, not aware of the distinction Owen by then was making between such "analogies" and recapitulation, was ecstatic: "so Owen actually believes in this view!!!!"[29] Around the first of January 1839, a few months after reading Owen's article, Darwin began reading Johannes Müller's *Handbuch der Physiologie des Menschen*. There he found that Müller held recapitulation to be a false doctrine and that the latest investigations, namely those of von Baer, had disclosed the "real law" of development. Müller's *Handbuch* was published in English translation under the title *Elements of Physiology* in two volumes (1838 and 1842), each of which contained a summary of von Baer's conception of development, the second volume treating it at some length. Darwin finished volume 1 in January 1840 and volume 2 in April 1842. In each volume he marked the embryological discussions and included the pages on which they appeared in his list of important references in the back. At the end of his notes on volume 2 he commented: "best abstract against metamorphosis [i.e., recapitulation] which I have seen."[30] When he wrote his "Sketch of 1842" his embryology was no longer recapitulationist, and in the "Essay of 1844" he cited Müller and Owen as authorities for his rejection of his old view:

> It has often been asserted that the higher animal in each class passes through the state of a lower animal; for instance, that the mammal amongst the vertebrata passes through the state of a fish: but Müller denies this, and affirms that the young mammal is at no time a fish, as does Owen assert that the embryonic jelly-fish is at no time a polype, but that mammal and fish, jelly-fish and polype pass through the same state; the mammal and jelly-fish being only further developed or changed.[31]

It is Owen's lectures of 1843 and the second volume of Müller (1842) that Darwin is referring to here, but from some of his embryological notes it is clear that the briefer discussion in the first volume of Müller had sufficed to persuade him to give up recapitulation.

Darwin recognized very quickly after reading Müller that the law of embryonic resemblance was potentially a very strong argument for genetic connection ("community in embryonic structure reveals community of descent," he said in the *Origin*).[32] But it did not appear to him that it was a necessary consequence of evolution by natural selection. If the descendants of some fishlike form have become adapted for life on land, why should their embryos still be fishlike? In January 1840, the month in which he recorded his reading of Müller's volume 1, Darwin began trying to answer this question:

> Selection only affects born individuals ((but their structure in variations depends on *foetus*)) (or eggs or larvae) & therefore only such are modified in relation to external world ((we know from diseases, that affections at one time of life are inheritable)) – such modifications will act directly to certain extent on foetuses in regard to size & proportion (as big-buttocked calves) of foetus with respect to bulk & quantity of nutriment &c. Hence foetuses in early stage or larvae would change far slower than full-grown animals, although undoubtedly there would be some change ((possibly only a virtual change)) when alteration had gone to a great extent by a kind of reflex action or vis medicatrix. – Hence we might expect that the foetuses & young & larvae (though in less degree, because I conceive larvae (silk worms!) might be *selected*) would show generic or ordinal affinities & be less useful in specific, at least between close species. – Hence also we might expect to find traces of parent form, long since dead, in foetuses – such being not injurious & being allowed to remain – of the same kind as rudimental or abortive structures, but more common & striking. – Hence we can understand teeth in whales & parrots bills – incomprehensible fact which on ordinary view of species must astound every reflecting mind. – Parrots must be less altered descendants of some form more intermediate between Mammalia ((?)) & present birds, than other birds – Hence first class. – [33]

In effect this first explanation of embryonic resemblances is that they can be seen to be a consequence of natural selection when one considers carefully the situations in which selection is able to operate. Natural selection can act only on "born individuals," those that in some sense have to struggle against other individuals and external conditions. Larvae, whose survival depends on adaptation to conditions, may be selected, but not fetuses and young protected by their parents. Young and fetuses will therefore have been less modified from the young and fetal conditions of their ancestors than will adult forms; and if they are members of the same group, so that they share a common ancestral state, they will much more closely resemble the common ancestor and each other than will adult forms. In larvae, on the other hand, such resemblances will have been to a greater extent obscured.

Darwin soon concluded that this first, simple hypothesis was not entirely adequate. It might be true that selection, which is the chief cause of modification, does not much affect embryos. But might not the variations on which selection works first appear in the embryo, and so change it? And might there not be other variations which only affect the embryo? Even if not selected, these would interfere with the law of embryonic resemblance. In a long note dated February 1841, Darwin attempted to deal with the second of these objections by saying that variations that tend to be hereditary at a young age will be obliterated by crossing, because selection does not act on the fetus.[34] At the same time he began working out a different explanation that would eliminate both objections at once. His new line of speculation drew on his theory of variation, rather than selection.

It was commonly believed, partly as a result of recent interest in teratology, that most hereditary variations were caused in the womb. Darwin's theory of variation, which said that external changes affect the reproductive systems of organisms, was in close agreement with this assumption. But Darwin's theory was flexible. When he began struggling with the law of embryonic resemblance, he considered and found some evidence for the possibility that variations, no matter when caused, may appear "at all periods of life . . . and that every change is not necessarily effect of some change or conditions in embryo – the embryo itself not necessarily affected, excepting that its future growth at different periods is to vary."[35] Together with Darwin's law of heredity, this would have the effect of exempting the embryo from some of the total number of variations that affect mature forms and so would leave the embryo less altered than the adult by the process of evolutionary change. Darwin believed that except when changes in external conditions cause variability, offspring resemble their parents. Stated more rigorously, so as to be useful in his embryological work, his assumption was that "each individual has tendency to beget another which shall be similar to itself at each corresponding time of life." The inheritance of variations, like the inheritance of firmly fixed characters, follows this law, he argued.[36] Therefore variations that first appear at some stage other than the embryonic will be inherited at that same stage and will not produce any alteration in the embryonic form of the species.

Variation does not always appear at an early period, he said in the "Essay of 1844." "It is at least quite *possible* for the primary germinal vesicle to be impressed with a tendency to produce some change on the growing tissues which will not be fully effected till the animal is advanced in life." And if the variations that over many generations have produced a new form first appeared at various times of life, they will, when added up by selection, produce their full complement of change only in the adult. The embryo will be affected only by those variations that first appeared in embryo, and so it will be proportionally less altered from the ancestral form.[37] Even before he wrote this, however, Darwin had reached a stronger conclusion, one that seemed to account still more satisfactorily for the fact of embryonic resemblance. It first appears in a passage published as a note in the "Sketch of 1842": "the parent more variable than foetus, which explains all."[38] The basis for this claim was Darwin's theory of variation. The adult is more variable than the fetus because it is more subject to the influence of external conditions than is the embryo. In a brief note apparently written soon after the "Essay of 1844," selection plays no part in the explanation of resemblances. All depends on this hypothesis about time of variation and inheritance at corresponding ages: "If all variations appear first generally not at earliest period of life (though ((cause)) potentially present in germ) & if they reappear at nearly corresponding ages, then the grand fact of similarity of embryonic structure is explained." These are Darwin's two embryological "principles," stated in nearly the form in which they appear in the *Origin*.[39]

In the "Sketch" and "Essay" of the 1840s Darwin's evidence for his various embryological arguments came mainly from the experience of breeders and the observations of medical men (for instance, heritable diseases that occur at the same period of life in successive generations). But when he presented his principles in the *Origin,* he was able to employ also research of his own which he undertook in the 1850s for the express purpose of validating them. In 1844 Darwin reported only a single comparison, of bulldog and greyhound puppies, to show that young differ less than mature animals, and he observed that it would be desirable to establish this on the basis of more numerous measurements as a general proposition for domestic animals. In the margin he made a note to himself, probably in 1854, to "get young pigeons."[40] He fitted up a pigeon house in the early spring of 1855 and soon was breeding birds of different varieties in order to kill, measure, and compare their young at different ages. "The chief points which I am, and have been for years, very curious about," he reported as he began, "is to ascertain whether the *young* of our domestic breeds differ as much from each other as do their parents, and I have no faith in anything short of actual measurement and the Rule of Three." The pigeon project proved to be difficult, time consuming, and initially distasteful to Darwin ("I have done the black deed and murdered an angelic little fantail and pouter at ten days old").[41] And like many other Darwin enterprises, it expanded and stretched over a number of years, eventually producing results that were relevant to several theoretical issues.[42] But its first fruit was that for which it was initiated. Normally the young birds were found not to have acquired the full amount of proportional difference that distinguished their parents.[43] The importance of this work to Darwin was not to establish the law of embryonic resemblance, but rather to show that it held good among a group of domestic animals which were known to have been produced by variation and selection. This allowed Darwin to interpret his results as evidence for his two principles of variation and inheritance. He could claim that the greater difference of the adult birds showed that the variations which man's selection has accumulated in order to produce the many breeds of pigeon "have not generally first appeared at an early period of life, and have been inherited by the offspring at a corresponding not early period."[44]

This, very briefly, is how Darwin's two principles developed from the time he read Müller until he presented them with the supporting evidence from pigeons in the *Origin*. But a few more words are in order concerning both their relationship to the main lines of Darwin's evolutionary thought and the nature of the problem they were intended to resolve. The two principles were not wholly new creations of the period after Darwin read Müller. They have their roots in his ideas about variation and heredity, which date from the period when Darwin's most fundamental belief about the organic world was that it is a harmonious system where all things are perfectly adapted. The principle of inheritance at corresponding ages followed fairly directly from his assumption that except when changing conditions interfere, offspring closely resemble their parents. The principle that variations occur usually at a not very early age he derived from his theory of the causes of variation, which gave him reason to think that variations will not be distributed evenly over all periods of life. In the early transmutation notebooks Darwin argued that variation occurs in order to adapt organisms to a changing world; after he read Malthus, and gave up the idea of adaptive variation, he continued to suppose that variations are caused by the changes in conditions to which the organism must adapt. Since he believed the main cause of variations was external changes, Darwin could explain why, as he then thought, internal parts are less variable than external. He extended the same reasoning to embryos: "Central organs bear some relation to embryonic in this respect," he said.[45] If external conditions can most easily affect external parts, they can also most easily affect organisms after they leave the egg or womb. Most variations should therefore appear at a not very early period of life. Obviously Darwin's principles, which required several years to formulate, were not inevitable conclusions from his theory of variation; but they were possible conclusions he was able to draw when confronted with the problem of embryonic resemblance.

While Darwin's solution to the problem is further evidence, in addition to that offered in Chapter 3, of the importance of pre-Malthus elements of his theory in his theoretical work during the 1840s, the problem itself is a further indication of Darwin's attitude toward the laws of his contemporaries. Darwin's self-imposed task was not merely to explain cases in which embryonic resemblances occur. It would have been easy to suppose that evolution might sometimes leave embryos relatively unchanged while at other times it changes them drastically. The cases in which resemblances occur would be evidence of common ancestry, to which Darwin could have applied one of his most frequently used arguments: Similarity of early embryonic stages is evidence of common descent because, except on a theory of descent, the fact is unintelligible, or is just an "ultimate fact." But von Baer had not discovered mere cases of embryonic resemblance. He had discovered a law, a regularity in nature, and this made things much more difficult for Darwin. We can say now that nature is not so regular as von Baer implied. And we can say that von Baer's law is in effect simply an expression of the conservative nature of heredity.[46] But this is not the way von Baer saw it, nor is it the way Darwin saw it. Darwin of course was required on morphological and embryological grounds to say that heredity is conservative; hence his principle of hereditary fixing. But in the case of the law of embryonic resemblance he believed he had also to show why there is a consistency and pattern to this conversatism. From von Baer's law Darwin concluded not that evolution will sometimes leave embryos unaltered, but rather that it will tend always to leave them unaltered. In order to bring his theory into harmony with von Baer's law he had to find an explanation for this unexpected tendency in the evolutionary process.

For a number of years after he began to speculate on this tendency toward the production of embryonic resemblances Darwin was unaware of the other side of von Baer's theory of development – that from the initially similar embryonic states there is an orderly divergence to the different adult forms. The law of embryonic divergence was brought forcefully to Darwin's attention only in December 1846, when he read Milne Edwards's "Considérations sur quelques Principes Relatif à la Classification Naturelle des Animaux." Darwin concluded from this that there must be in the evolutionary process another tendency, which he also had not expected, to make embryos diverge from one another according to the degree of modification of the adult forms. He immediately began a series of speculative notes on the subject that continued until at least 1858, without however leading to any explanation that he was willing to publish. Darwin's work on embryonic divergence is noteworthy not because of its results but because it reveals a shift in his thinking about the branching character of evolutionary change.

Milne Edwards's essay was an attempt to apply von Baer's embryological ideas to the problems of classification. This had been suggested by von Baer himself, but Milne Edwards carried it out in a more complete and thorough manner. In the first half of the article he presented an outline of his views on embryology and his reasons for concluding that classification ought to be based on embryonic characters.[47] In the second half he gave a summary view of the classification of the vertebrates, particularly the mammals, to demonstrate the utility of his principles. While reading the second half, where Milne Edwards's theoretical statements are supported with concrete examples, Darwin expressed surprise at the law – which he interpreted, as usual, in evolutionary terms – that there is a correspondence between the period of embryonic divergence and the affinity of the mature animals:

> Fish & Batrachians, continue for long to walk in the same embryonic route: whereas Mammals, Birds & Reptiles, at an earlier date begin to diverge or "la marche génésique parallèle (in these) est de moins longue duré" [This is something odd: if you alter an organ considerably, you alter the embryo up to a certain early age; if you alter it still more you alter it earlier still & so on . . .][48]

Milne Edwards's statement that groups that differ more diverge sooner during development than those that are more similar is taken directly from von Baer: "*the more different two animal forms are, so much the further back in development must one go to find a correspondence.*"[49] This, von Baer said, is merely a "changed form of expression" of his basic law that the more general type appears before the more special. The developing chick is at first merely a vertebrate, then a land vertebrate, then a bird, and so on; which is to say that it diverges first from fish and amphibians, then from reptiles and mammals, until it acquires the general characters of its class. It may well be asked then why, after Darwin had read Müller's account of von Baer's views, this should strike him as odd?

To begin with, Müller may be said to have understated the importance of divergence in von Baer's theory. He could not discuss von Baer's views without implying that there is divergence, but he did not insist on the regularity of the phenomenon, as von Baer and Milne Edwards did. Müller saw in von Baer primarily proof that vertebrates can never recapitulate the forms of animals of the other *embranchements*. He was impressed also by the fact that all vertebrates are alike at first but then develop in different directions, the initially similar limb buds becoming fins, wings, feet, hands, and so forth. Man therefore does not repeat the forms of the lower vertebrates, which all depart in different ways and to different degrees from the common path of development. It is only here that divergence enters Müller's discussion. The subject is embryology, and there is no reference to the implications von Baer saw in his theory for the study of affinities, no general statement of the law of divergence, such as that animals first acquire the form of their type, then that of their class, order,

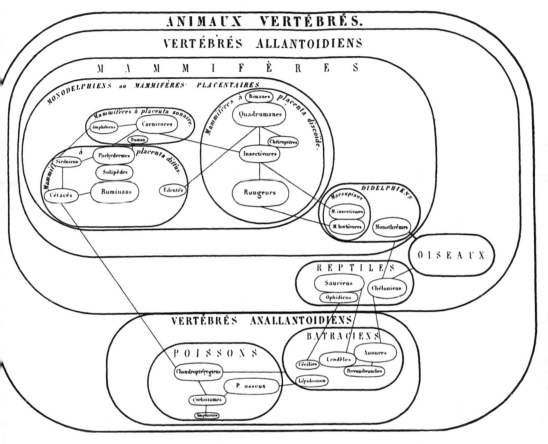

Milne Edwards's embryological classification of vertebrates. ("Considérations sur Quelques Principes Relatifs A la Classification Naturelle des Animaux," *Annales des Sciences Naturelles*, 3rd ser., 1 [1844]; foll. p. 98.)

family, and so on. From reading Müller a recapitulationist might be induced to abandon recapitulation and yet retain a strong linear component in his conception of development. Certainly he would not be forced to conceive of embryonic divergence as a regular phenomenon whose timing is closely correlated with the degrees of relationship among the adult members of each group of animals.[50]

In the "Sketch of 1842" and the "Essay of 1844" Darwin's treatment of embryology assumes only the bare minimum of divergence. The mammal is at no time a fish, the jellyfish at no time a polype. The fish must therefore at some period diverge from the path followed by the mammalian embryo in order to assume its own fish form, and the polype similarly with respect to the jelly fish. On the basis of these statements it is safe to place Darwin among the followers of von Baer, rather than of the recapitulationists. But it appears safe to say also that embryonic divergence was less important for Darwin than for von Baer. Darwin's chief interest was in the resemblance of early embryonic stages. When he referred to the differences that development ultimately produces between adult fish and mammal, he attributed them principally to the higher development of the mammal, rather than to divergence.[51] In this his emphasis was the opposite of von Baer's. Von Baer agreed that the embryonic mammal will resemble the adult fish because the fish is not far removed from the vertebrate type, that is, not highly developed. But for him the fundamental differences between fish and mammal were due to their assuming different subordinate types of structure, not to their achieving different degrees of development.

Other sources besides the "Sketch" and "Essay" seem also to suggest that before Darwin read Milne Edwards embryonic divergence was relatively unimportant to him and that he continued to think of development as a more or less linear progress, a remnant perhaps of his former belief in recapitulation. Consider for instance this note from November 1844, in which Darwin worked out his explanation for one class of exceptions to the law of embryonic resemblances – those animals whose young resemble their own parents rather than the embryos of related forms:

> In Embryology if embryo passes through several successive stages, then my explanation requires that the embryonic stage shall have become ((probably)) longer & longer ((& more complicated)) – for by this, the selection only materially altering the form after birth, will allow a succession of forms to be passed through before birth. If embryonic period be shortened, say if mammal born under fish form, then the fish form would have to be altered to conditions of life, & thus reptile form might be lost. If born in shorter time & then again in shorter, the mammal form w^d encroach on the reptile & then on the fish-form, & the transition would have to be more & more rapid till lost, *for selection might fall* on the primary vesicle & every

successive transitional stage be absolutely lost & destroyed – [applica-
ble to Spiders & Crabs][52]

This is the germ of Darwin's explanation in the *Origin* of spiders
and cephalopods, two groups of animals that Owen in his lectures
had cited as undergoing no metamorphosis.[53] The principle
interest of the note, however, lies in the model of development
Darwin seems to have been employing. He considered the changes
the embryo undergoes as a series of stages that reflect in some way
the ancestry of the individual. Because mammals have descended
from fish and reptiles, their embryos pass through a "fish form"
and a "reptile form," which have been added onto the originally
shorter developmental path of some ancestral vertebrate. These
stages tend to be preserved during the evolutionary process
because selection only alters the form after birth. But stages that
have been added may later be deleted, just as in Darwin's old view
of generation as a "shortened" repetition of ancestral forms. If we
had no other record of Darwin's embryological ideas in the 1840s,
we would surely conclude from this document that he was a
recapitulationist and that he was accounting for an exception not
to the law of embryonic resemblance but to the theory of
recapitulation. Since we do have other records we can recognize
that by "fish form" Darwin meant not an adult fish but the stage at
which mammals and fish are still similar, before the fish diverges
to assume its special mature form. Nevertheless, it is difficult to
deny the predominance of the linear component over the divergent in Darwin's embryological thought at this period.

From the time he read Müller until he read Milne Edwards,
Darwin's embryological speculations were almost exclusively concerned with the question of embryonic resemblances – their
explanation, their significance for classification and paleontology,
exceptions to the rule. The conclusion toward which these speculations pointed was that the evolutionary process tends to leave
embryos unaltered. In what appears to be the earliest unambigu-
ous statement of his two principles (part of which has already been
quoted), Darwin recognized the logical consequences of his
reasoning up to that point:

If all variations appear first generally not at earliest period of life
(though ((cause)) potentially present in germ) & if they reappear at
nearly corresponding ages, then the grand face of similarity of
embryonic structure is explained – *If the proposition absolute embryo w^d*

seem identical, but no doubt some effect is produced in embryo. If any appeared early, then, embryo more affected – Even direct causes of variation w^d not affect embryo.[54]

It is tempting to place the emphasis on the disclaimer ("no doubt some effect is produced") and say that of course Darwin realized that the evolutionary process would not leave embryos utterly unchanged. But it is well to recall that in 1844 Darwin still believed that there is scarcely any variation among organisms in a state of nature. Individuals of a species are like objects cast in the same mold; only when they are subjected to changed conditions of life do they vary. One who believed in this sort of constancy could well conceive of the nearly absolute identity of embryos from generation to generation, since they, unlike "born individuals," are isolated from changes in conditions.

Milne Edwards's essay, by contrast, confronted Darwin with a law that from his evolutionary perspective indicated that there is a strict and regular correspondence between embryonic change and the evolutionary modification of the adult.[55] Milne Edwards presented divergence as a general phenomenon in embryonic development and in the order of nature as revealed by taxonomy. To Darwin this implied the existence of tendencies in the evolutionary process that required explanation.

For some years Darwin thought that the tendency toward embryonic divergence could be accounted for by a "law" of development proposed by Auguste Brullé. Darwin's evolutionary interpretation of Brullé's law was that the more a part is modified by natural selection, the earlier it begins to be differentiated in the embryo.[56] This, he observed to Huxley in 1857, would explain the strict relationship Milne Edwards proposed between embryonic divergence and degrees of affinity:

> Especially I want your opinion how far you think I am right in bringing in Milne Edwards view of classification. I was long ago struck with the principle referred to: but I could then see no rational explanation why affinities should go with the *more or less early* branching off from a common embryonic form. But if MM Brullé and Barneoud are right, it seems to me we get some light on Milne Edwards views of classification; and this particularly interests me.[57]

Huxley responded that Brullé's law was wrong, and this set Darwin to work on a new series of speculations, which were still incomplete when he was interrupted by Wallace's letter in June 1858.[58]

Though it did not lead directly to any change in his theory, Darwin's perception of the law of embryonic divergence as a problem for him is in some ways more interesting than his similar reaction to the law of embryonic resemblance. It indicates that consideration of Milne Edwards's article on classification led Darwin to see divergence as a more regular and general phenomenon than he had previously supposed it was. I will suggest in the following chapter that this contributed to the initiation of a major new research project on the causes of divergence generally.

The evidence for descent

The part of his embryological work that he published, chiefly his two principles, Darwin reckoned among his finest achievements. Embryology was his "pet bit" in the *Origin*, he said, and the law of embryonic resemblance was in his view "by far the strongest single class of facts in favour of change of forms." He considered the "morphological or homological argument" a close second to the embryological in its conclusiveness as evidence of descent, and in his correspondence he often linked the two, as in this letter of 1860:

> I rather doubt whether you see how far, as it seems to me, the argument for homology and embryology may be carried. I do not look at this as mere analogy. I would as soon believe that fossil shells were mere mockeries of real shells as that the same bones in the foot of a dog and wing of a bat, or the similar embryo of mammal and bird, had not a direct signification, and that the signification can be unity of descent or nothing.[59]

In concluding the chapter of the *Origin* on the various morphological departments of natural history (Chapter XIII), Darwin said that the several classes of facts presented in it were such a clear proof of common descent that he would "without hesitation adopt this view, even if it were unsupported by other facts or arguments."[60]

Darwin evidently judged that the other Part II subjects, paleontology and geographical distribution, were less persuasive, although he did occasionally refer to them as providing valuable support for his theory.[61] Because of its important role in leading Darwin to become a transmutationist and its contribution to the formulation of the theory of natural selection, geographical distribution has received a great deal of attention from Darwin scholars.[62] But paleontology has not, which is somewhat surprising, for it was above all paleontology that raised the issue of descent and made it an attractive hypothesis in the mid-nineteenth century. Paleontology did not supply detailed evidence for change of the kind Darwin would have liked – finely graduated series from species to species – and it is perhaps for this reason that the contributions of pre-1859 paleontology to the establishment of evolution were, until the publication of Peter Bowler's book, pretty much ignored.[63] But Darwin nevertheless benefited considerably from the mid-century revision of Cuvier's account of the geological succession of organisms. As reinterpreted by Owen, Carpenter, and others, the fossil record gave a developmental picture of the history of life that supported the idea of a genetic connection among successively appearing forms. There was a wealth of new fossil material discovered during the years Darwin was at work on his theory, and his paleontological notes are full of references to species that were described – most often by Owen – in terms that suggested branching development. Darwin was especially excited by Owen's reclassification of Cuvier's pachyderms and ruminants, because extinct forms were the basis for Owen's rearranging and uniting into two related series groups that Cuvier had treated as among the most distinct orders of mammals.[64]

While new fossils were important for Darwin, he was even more fortunate in the new generalizations formulated by contemporary paleontologists.[65] For instance, Darwin's belief in recapitulation led him at an early period to suppose there should be a parallel between embryological and paleontological development.[66] When Agassiz proposed this as a law of nature, Darwin naturally cited it in support of his views. His embryological principles let him explain why the embryo should be left unchanged and so resemble the *embryo* of the ancestral form; and he saw the resemblances Agassiz, Owen, and others noted between modern embryos and *adult* fossil forms as evidence for the kind of morphological relationship he expected to find between ancient and modern forms. Darwin interpreted the parallel not, like Agassiz, as recapitulation, but like Owen and Carpenter, as development from the common archetype or ancestor. The parallel should exist, he thought, because ancestors were in his view relatively undifferentiated, archetypal forms.[67] The importance of the parallel to Darwin as an argument for evolution is perhaps best indicated by the fact that it figured prominently in his first brief exposition of his theory to Lyell, whose support he very much wanted. In the *Origin* he alluded to Huxley's criticisms of "Agassiz's law," but he expressed confidence for some groups.[68]

In the *Origin* Darwin cited also the law that "the more ancient a form is, by so much the more it tends to connect by some of its characters groups now widely separated from each other."[69] Some of the older writers, such as Buckland, had occasionally noticed the transitional character of many fossil forms. But by the early 1850s the fact that extinct forms are frequently intermediate between modern groups was asserted by numerous competent paleontologists who did not mean, as Buckland had, that they filled gaps in the chain of being.[70] In Carpenter's discussion of 1854 of the paleontological application of von Baer's law, such forms were treated as one kind of evidence for the generally developmental picture he gave of the history of life.[71] This newly established view of the fossil record, together with the new fossil evidence that was being interpreted in accordance with it, was sufficiently supportive of Darwin's argument that he could readily dismiss the few critics (particularly Huxley) who said there were no intermediate forms and no divergence and confidently offer branching evolution as the best explanation of the paleontology of his day. Apart from the imperfection of the geological record, he said, "all the other great leading facts in paleontology seem to me simply to follow on the theory of descent with modification through natural selection."[72] This is a claim that could not well have been made in 1838, because most of the relevant "great leading facts" he had in mind had not then been established.

Despite its relative brevity in the *Origin*, Darwin's treatment of the subjects forming Part II of his species work must be counted a considerable achievement. He took a major portion of contemporary natural history and made of it an impressive argument for the fact of descent. For homologies and unity of type the argument was straightforward. Paleontology required slightly more finesse, for the deficiencies of the evidence made it necessary for Darwin to explain rather carefully what one ought to expect on the basis of geological knowledge and of his theory – that there will be gaps in the record, that intermediate forms will be rare, and what intermediate forms will be like. But once this was done the generalizations of paleontologists could be enlisted in support of descent. Embryology posed special problems of another sort. Because he thought he had to explain tendencies of nature, Darwin was forced to add to his theory to bring it into harmony with von Baer's laws. (In the next chapter I will discuss a similar problem he faced with respect to classification.)

Darwin frequently complained that his critics paid too little attention to the arguments from morphology and embryology.[73] One reason for their comparative neglect is not far to seek. Much of Darwin's reinterpretation was self-evident and consequently noncontroversial.[74] Even the facts of embryonic resemblance and divergence, which caused him such problems, must have seemed to most biologists to be obviously explicable by descent if descent were true. Well-informed readers in the scientific community were becoming increasingly well aware that descent (though perhaps not transmutation) could explain the leading facts in mid-century morphology, paleontology, and embryology, and for the nonprofessional, Chambers's *Vestiges* had pointed to the same conclusion.[75] But though it was noncontroversial, Darwin's recasting of natural history was undoubtedly effective. It is often suggested, by Darwin himself for instance, that descent was rapidly accepted after the publication of the *Origin* because Darwin presented a superior theory to account for it.[76] Descent was an old idea; natural selection was the novelty in Darwin's book. Darwin's theory did probably contribute to the change of opinion that occurred after 1859, if only because it showed the possibility of a simple explanation of organic change. But the paradox noted by Ellegård, that it was descent, not natural selection, that was accepted,[77] is less bothersome if one recognizes that natural selection was not in fact the only novelty in the *Origin*. Part II was new as well. Darwin was the first reputable naturalist to argue for transmutation who was able to take advantage of the developmental view of nature that emerged in the mid-nineteenth century. This, as much as his new theory, separates Darwin's work from Lamarck's. Darwin showed, point by point, how a professional biologist, rather than a popularizer, would explain the latest generalizations of his science by the hypothesis of descent. In doing so he demonstrated even to the well-prepared reader just how strong the case for evolution could be made. The paleontologist H. G. Bronn commented that although the theory of evolution is not new, "it appears to be stated by Darwin with all the intelligence and knowledge which only the present state of science affords the talented investigator."[78]

It is arguable that among that small but important group, professional biologists, Darwin succeeded in large measure because he did his work well in Part II and because contemporary natural history gave him good material with which to do it. Since I have not here undertaken a study of the reception of the *Origin* by professional biologists, this must remain a suggestion. What is clear, however, is that Darwin thought the success of his theory would depend on his evolutionary transformation of the science of his colleagues. This he asserted again and again in letters to friends and critics.[79] To repeat what I said earlier, Darwin's recasting of contemporary biology ought not to be passed over lightly. The "classes of facts" he explained had not been lying around for ages, their significance unrecognized until he came along to reveal their meaning. They were for the most part new generalizations. It is for this reason that the problems Darwin worked on, his vision of the shape of the evolving organic world, the arguments he was able to offer in support of his theory,[80] and the rapid acceptance of evolution after 1859 must all be viewed in the context of the recent theoretical work of his fellow naturalists.[81]

7

Classification and the principle of divergence

On September 9, 1854, as he was packing up his specimens and reading proof for his last volume on cirripedes, Darwin entered in his *Journal* that he "began sorting notes for Species theory." From then on, he recalled in his *Autobiography*, "I devoted all my time to arranging my huge pile of notes, to observing, and experimenting, in relation to the transmutation of species."[1] This brief catalogue of activities ought also to include "theorizing," for September 1854 marks the beginning of one of the most fruitful periods in the development of Darwin's evolutionary thought.

Darwin began by reading over his "Essay of 1844" and his collection of notes, writing memoranda to himself about the evidence he needed to collect on various topics and about problems that were still to be resolved. Most of the matters he noted as requiring attention pertained to Part II of his work, and over the next four years his observing, experimenting, and data gathering centered on Part II subjects: embryology (measurements on young pigeons), geographical distribution (experiments on transport of seeds), paleontology and extinction (compilation of data on aberrant forms), and classification (experiments on the lawn at Down, compilation of data on variation in large genera and related subjects). Darwin's envelopes of notes for Part II indicate that the major problem he saw in the fall of 1854, the one to which he first turned his attention, was how to explain the branching, treelike appearance of current schemes of classifying organisms. Apart from the writing of *Natural Selection,* solving this problem and constructing an argument to support his solution became Darwin's largest single project – gauged by the amount of his own time spent on it and by the number of assistants he recruited – during the period 1854–8.[2] The result of this work was Darwin's "principle of divergence," the most important addition to his theory between 1838 and 1959 and the one most intimately associated with the transformation of his theory after 1844.

The problem of classification

Because the principle of divergence is included in the chapter on natural selection in both *Natural Selection* and the *Origin,* it has sometimes been assumed that the problem it was designed to solve was how to account for speciation: How do varying offspring of one species give rise to two or more species? Or how are the slight differences between varieties augmented into the greater differences between species?[3] These are questions to which the principle of divergence was ultimately applied, but they were only stated in this form when Darwin realized that his new principle gave an answer to them. All Darwin's notes indicate that the "one problem of great importance" that he overlooked until after 1844 concerned classification, not the multiplication of species.[4] The question he posed was not how one parent species gives rise to two descendant species, but rather "why the species of a large genus, will hereafter probably be a Family with several genera." It is a trivial point, but worth noting, that the envelope of scraps that was later labeled "loose collected notes on the principle of divergence" turns out on inspection to be Darwin's file of notes on classification, only some of which have anything to do with divergence.[5]

The whole train of Darwin's speculations that I am about to discuss provides much additional evidence that the principle of divergence grew out of work on classification. The question this raises is what exactly was the problem, and when did Darwin come to see it as a problem? From 1837 on Darwin assumed that the natural system of classification should be genealogical and that lines of affinity may be represented by the "tree" or "coral" of life.[6] However it was several years before he concluded that there must be a tendency to diverge to account for this branching arrangement. When he discussed "affinities and classification" in the "Essay of 1844," Darwin employed only two interrelated factors, both of which were already part of his theory of descent, to explain phenomena he would later attribute to the action of divergence. The fact that organisms fall into groups within groups follows, he said, from the descent of whole genera and families from the offspring of a single species; and the very existence of groups is due to extinction, which produces the gaps that serve to delineate the boundaries of groups at all levels. If a species spreads into six separate regions, or if it is widely diffused over an area that is then divided into six regions, and if the conditions in these regions differ, the species, not being perfectly adapted to the various conditions it encounters, will vary in each region. Natural selection might then produce six new species, which will either form a new genus or be ranked as one section of the genus to which the parent species belonged. Subsequent changes will probably cause the extinction of some of the six, but if two or three survive, they, from inheriting the advantages that let their parent spread, will likely spread also and give rise in the same manner to two or three small groups of species. These groups will be more or less similar, depending on how different were the two or three species that spawned them. They might form several sections of one genus or several new genera. If the latter, these together would be a family. Thus the spreading of a successful species and its descendants into different regions produces a family that contains numerous genera and many more species, group within group within group. But without extinction there would be no separate groups whatever: "the arrangement of species in groups is due to partial extinction." Were there no extinction (or, more properly, no forms that have not left remains, either unaltered descendants or fossils), the descendants of the original species would constitute a single large group in which no subordinate groups could be distinguished because all the forms would be separated only by small gradations.[7]

In 1844 the classification of organisms posed no special problems for Darwin, and he required no special principles to account for it. In the "Essay" classification follows naturally from his theory of descent. "All the leading facts in the affinities and classification of organic beings can be explained on the theory of the natural system being simply a genealogical one," he said.[8] Contrast this with his remark to Hooker in 1859: "[Naudin's] simile of tree and classification is like mine (and others), but he cannot, I think, have reflected much on the subject, otherwise he would see that genealogy by itself does not give classification."[9] I have found no single statement by Darwin that indicates what led him to this conclusion and induced him to begin to look for a principle of divergence, but several taken together provide an answer. Between 1844 and 1847 Darwin became increasingly aware that the leading biologists of his generation had begun to think of organic nature as branching upward; and Milne Edwards's work on classification in particular indicated to him that divergence could not be the occasional and accidental result of the diffusion of single species that he described in the "Essay of 1844," but must be rather a universal tendency among organisms.

That Darwin had changed his mind by 1847 is apparent from the following note, written in July of that year, which suggests that he had begun to look at the production of the branching arrangement as a problem to be resolved:

> The affinities of organisms are represented by distance – species being called dots . by their being placed thus
>
> . . .   .   •••••••• • ••• . • . • •••   • • • •   • • •   ••• •   • • • • •
>
> As this arrangement leads to idea of common genetic causation, as we see with chemists in proportion of elements, we are led to compare things thus arranged to branching of tree & may be said to diverge from common stem. –
>
> Begin with stating fact – universal at all times & places – overlooked for familiarity, – genus not an entity – Begin with single species vars – action of divergence – the same cause which formed several species will make others & when once formed into larger genus, we know these are the very groups, which do vary most & therefore will give rise to more varieties & thus give divergent character. – Here again we have resemblance to most great forest trees in which 2 or 3 great branches from some accidental advantage have exterminated the others. These affinities are represented down the lines of stem.[10]

The assumption here is that some cause other than mere accidental dispersal and natural selection is required to produce divergence, but besides the vague reference to "action of divergence" there is little indication of what that cause might be.[11]

Two notes from 1856, when Darwin had found such a cause and was drafting tentative paragraphs to explain it, point to the recent shift away from teleological explanation in natural history as one source of his problem. As we have seen, most mid-century biologists no longer believed that organisms were constructed solely with reference to their conditions of existence. Instead they supposed them to be formed on various types, subtypes, subsubtypes, and so on, which implies a branching classification.[12] This required Darwin to show that descent would necessarily produce a branching arrangement in each group of organisms. It was all the more necessary to do so because nonevolutionary explanations were available, such as Owen's archetypes and Milne Edwards's laws of variety and economy. In December 1856 Darwin wrote, "What are called important parts vary seldom. – & so they differ only seldom in great group – now it is probable that most diverse are apt to be propagated; if so it would ensure that we sh^d have type with important differences [that is, subtypes]."[13] More explicit is this longer note dated May 11, 1856, and headed "Classification":

> When a new species is to be created, it might naturally be supposed that a new form of those groups which were best adapted w^d be created; & this w^d be perfect explanation, if the type on which the new form was to be created had its whole organization simply related to conditions.[14] But few admit this [is the] case, that skull is formed from vertebra owing to mere adaptation . . . But it is confessed that it is adaptation in accordance with an ((preexistent)) idea, plan, or archetype. Therefore new species must be created to some preexistent idea or plan. –
> But this, so satisfactory to many eminent men, to my mind is only saying that so it is. – how according to descent can be explained this curious arrangement of all living & extinct beings.[15]

Clearly Darwin wanted to provide an alternative to archetypes and subtypes. But why did he think the theory of common descent alone was insufficient?

If we turn back to Darwin's reading in the years 1844–7 we can see why it was then that classification became a problem for him and why he concluded that he required an additional explanatory principle to account for branching. It was precisely in this period that branching began to appear as a dominant theme in the work of several naturalists Darwin respected. Darwin noted in the *Athenaeum* in 1845 that Edward Forbes in a lecture at the Royal Institution described affinities as branching off low down in each series.[16] In March 1846 he visited Owen, who gave him a "grand lecture on a cod's head" and explained his views on the vertebrate archetype.[17] Later that year Owen's "Report on the Archetype" appeared in the *Report* of the British Association for 1846.[18] In December of 1846 Darwin read Milne Edwards's "Considérations sur quelques Principes Relatifs à la Classification Naturelle des Animaux," which he called "the most profound paper I have ever seen on affinities."[19] Given the nature of Darwin's project in Part II, the growing consensus that the natural arrangement resembles a branching tree required him to give some explanation of the fact. As long as Darwin saw the idea of branching as peculiar to his own view of nature, he could freely use it as a tool to explain phenomena that others looked at differently, such as the arrangement of organisms group within group. But when others who did not necessarily believe in descent explicitly connected group-in-group classification with the idea of a branching development of life, branching itself entered the category of contemporary laws demanding an evolutionary interpretation.

Milne Edwards's essay seemed to imply further that the construction of an evolutionary interpretation of branching called for something more than the statement that the members of every group are genetically related. Nowhere in the literature of mid-century morphology is the phenomenon of divergence more strikingly presented than in Milne Edwards's essay. Milne Edwards insisted on von Baer's law of embryonic divergence from common forms, and he persuasively argued that there is a direct parallel between embryonic divergence and the affinities of adults. The unmistakable message of his essay was that divergence is a general feature of nature. Viewed from Darwin's perspective such a generalization looked like the definition of a tendency in the evolutionary process, and it posed the same sort of problem for him as von Baer's laws of embryonic development. Darwin's explanation of classification in 1844 depended on the accidents of transport, geographical isolation, and extinction. It could explain how the offspring of some one species might spread and give rise to a genus or a family, but it could not explain why there should be a tendency of offspring to depart in different directions from the parent form. It could not show that nondivergent evolution was not the general rule. If divergence is of general occurrence, it requires an equally general explanation. "Mere chance," Darwin said in the *Origin,* could never account for it.[20]

Darwin's conviction of the generality of the phenomenon of divergence and the consequent necessity of providing an explanation for it was reinforced by his work on cirripedes. In December 1848, two years after reading Milne Edwards, he wrote:

> I have been much struck in Anatifera how the genus, (& I have no doubt universal, as evidenced by sub-genera) breaks up into little

groups. – hence those who use Diagnostic characters have generally
to refer to only 2 or 3 species – So again species break up into groups
of varieties. Genera again in same family are united into little groups
– so throughout animal kingdom. – so children even in same family –
It is universal law. – [21]

It was to explain this universal law that Darwin developed his
principle of divergence. The task he set himself when he resumed
species work in 1854 was to "show that there is a tendency to
diverge (if it may be so expressed) in offspring of every class; & so
to give the diverging tree-like appearance to the natural genealogy
of the organised world."[22]

The solution to the problem

In November 1854 Darwin recorded the following hypothesis in
his notes on classification:

> To explain why the species of a large genus, will hereafter probably
> be a Family with several genera, we must consider, that the species
> are widely spread, & therefore exposed to many conditions & several
> aggregations of species: they will occasionally mingle with a new
> group, & then on the principle, that the most diverse forms can best
> succeed, it may be selected to fill some new office, & mere chance w^d
> determine the origin in a large genus [of] some new & good
> modification.[23]

The basic assumption here is the same as in Darwin's account of
the origin of genera and families in the "Essay of 1844": Groups of
higher value are formed from species and genera which because
of some advantage are able to spread into new areas; there they
encounter new conditions that cause them to vary and give rise to
new forms. What is novel in this hypothesis of 1854 is the idea that
the varying offspring of spreading forms will tend to be selected.
They will have a better than average chance of survival because as
newcomers they will differ from the natives and so may fill some
unoccupied place in the local economy.

As early as the note of July 1847 quoted above, Darwin
suggested that large genera vary more and give rise to more
varieties. The new principle now introduced, that the most diverse
can best succeed, explains why there will be a tendency for these
large and successful groups to go on spreading, growing, and
diverging ever further from the parent stock. This is not yet the

principle of divergence that is presented in *Natural Selection* and the *Origin*. But as the idea that the most diverse can best succeed was elaborated over the next two years and integrated into the theory of natural selection, the principle of divergence emerged. The difference between the principle of divergence and the hypothesis of November 1854 is that in 1854 Darwin still assumed, as in 1844, that it was necessary for species to spread and encounter new conditions before they would vary and give rise to new forms, whereas according to the principle of divergence, the production of diverse offspring from one parent form may occur in a single locality without any change in conditions.[24]

As was his custom, Darwin turned to the study of geographical distribution to find support for his hypothesis. On November 15, 1854, he wrote Hooker that he had been trying to get "a general view" of the geographical distribution of animals.[25] His principal source was the entomologist C. J. Schoenherr's *Genera et Species Curculionidum,* in which he made a number of calculations in order to determine whether large genera in fact have wider ranges than small genera.[26] This was crucial to his tentative explanation of classification because of his assumption that groups grow large by spreading, being subjected to new conditions, and varying.[27] Darwin summarized his results in a note headed "Theoretical Geograph. Distrib," in which he showed that small genera tend to be "local" and that larger genera are indeed "widely extended," which was consistent with his hypothesis.[28] But although these calculations established the possibility that the hypothesis was right, they could provide no direct evidence for it. To show that large genera tend to become families and that they do so because the most diverse forms can best succeed required Darwin to produce some evidence that large genera are increasing; and it required him to demonstrate the validity of the principle that the most diverse can best succeed.

The first of these requirements posed a particularly difficult problem. How could one show that large genera are growing genera? Darwin's work on aberrant forms in the period November 1854 – March 1855 suggested one test that could be applied. His research in Schoenherr led him to begin an extensive series of calculations in search of evidence for his argument, developed years earlier, that aberrant genera, such as *Ornithorhynchus,* are not oddities created to form links between typical groups, but are rather remnants, "living fossils," which appear odd only because the genera most closely allied to them are extinct. Darwin reasoned that if aberrant genera are made so by extinction, then they are likely to have – as MacLeay and others said they did have – few species.[29] The fact of their aberrance shows that these genera belong to families and orders that are "less adapted" to the present condition of the world, and so it should follow that the species within aberrant genera will have suffered much extinction from the same cause, being "less adapted."[30] Although Darwin's calculations on the ranges of and numbers of species in aberrant genera did not provide as conclusive evidence as he had hoped for their being living fossils, his speculations on aberrants did lead to one argument for the proposition that large genera are increasing.[31] Using his long held (but soon to be abandoned) belief that the number of species on earth is approximately constant, he concluded that the fact of extinction "near & around the aberrant genera" implies that "creation has fallen on the typical & larger genera."[32]

The work on aberrants also suggested a criterion by which one could judge whether a group was a site of creation or extinction. At an early stage in his speculations on transmutation Darwin decided that if his theory were true then geographical regions in which species formation is in progress should be characterized by "fine" or "transitional" species. A new and rising group of islands, for instance, should have species closely related to each other and to their parent forms on the nearest continent.[33] Consideration of aberrant genera, whose species Darwin thought should be distinct because extinction has thinned their ranks, suggested by contrast that "close" or "fine" species should be characteristic not only of regions, but also of genera in which species formation is in progress: Aberrant or declining genera will have very distinct species, while increasing genera will have close species.[34] When in the spring of 1855 Darwin read, or reread, his notes on an article by Elias Fries, he saw that close species might provide just the evidence he needed to show that large genera are increasing and may therefore be expected to become a family. He had believed since the first transmutation notebook that the large genera are the increasing ones.[35] His work on aberrant genera led to the conclusion that increasing genera should have close species. And now he found in Fries the claim that large genera do have close species.[36]

This induced Darwin to begin a new series of calculations to determine whether there are more close species in large than in small genera. For this he required the assistance of naturalists who could tell him which were the close species in their own special areas of study. By the summer of 1855 he was requesting such assistance from Asa Gray, H. C. Watson, and his old teacher J. S. Henslow. The results, reported in *Natural Selection,* appeared to indicate a slight preponderance of close species in the larger genera.[37] But this was just the beginning of a much larger project. Fries gave Darwin the idea of trying to prove statistically that large genera have more varieties than small. Since by Darwin's theory varieties, "fine" species, and well-marked species form a continuum, he concluded from his close-species hypothesis that he should find not only more close species, but also more varieties in the large genera than in the small.[38] Here was another testable proposition that might support his explanation of classification. In the fall and winter of 1855–6 he went through numerous catalogues, mostly botanical, compiling statistics on the numbers of varieties in large and small genera as well as on the number of varieties in genera with and without close species and on the sizes of genera that include very common species.[39] The results were true to Darwin's expectations, and by the spring of 1856 his calculations had given him confidence that he could prove that large genera are usually increasing genera.

The second task imposed by his hypothesis was to establish the principle that the most diverse forms can best succeed. Ten years earlier, in a different context, Darwin had employed the same idea to explain some peculiar facts of geographical distribution. He observed that coral islets and extreme arctic regions have few species of plants but that these belong to many genera and orders. He thought that the explanation, at least for the arctic case, might be that "plants of diverse groups c^d subsist in greater numbers, & interfere less with each other."[40] Now he returned to the flora of coral islets as evidence for his principle. In *Natural Selection* he described the diverse flora of Keeling Atoll, collected by himself, and argued that such diversity as it displayed could only be explained by supposing that of all the plants whose seeds have probably been carried there, those alone that "differed greatly from the earlier occupants, were able to come into competition with them & so lay hold of the ground & survive."[41] But Darwin considered such cases to be too peculiar to prove that the principle was a general rule governing the economy of nature. "It is indispensable," he wrote in November 1854, "to show that in small & uniform area there are many Families & genera. For otherwise we cannot show that there is a tendency to diverge . . . in offspring of every class."[42] The following spring found Darwin and the governess collecting plants in a nearby field for this purpose. The fruits of their activity, along with the results of some small experiments on the Darwin lawn and other facts Darwin gleaned from his reading, are offered in *Natural Selection* in support of his claim that diversity is beneficial.[43]

In January 1855 Darwin outlined another related argument. If diversity is an advantage, then where the inhabitants are much diversified, we should find that more life is supported. He proposed that the best measure of this would be the amount of chemical change occurring, as indicated by the "amount of carbonic acid expired or oxygen in plants." He made a similar suggestion in *Natural Selection,* but clearly an experiment to compare the amount of oxygen expired by the plants in equal areas of heath and meadow was out of the question for Darwin (or anyone else, for that matter).[44] He was fortunate, however, that a discussion on methods of sowing in the *Gardener's Chronicle* in 1856–7 brought to his notice some simpler experiments that others had already performed. These seemed to prove that as the number of species sown in a given area increases, so does the yield, which Darwin took to be good evidence for his principle.[45]

While Darwin was collecting evidence for the various parts of his hypothesis, he was also considering the relationship of his principle of the advantage of diversity to the theory of natural selection. He had initially assumed simply that the fact of their being different gave an advantage to the varying offspring of species that range widely. But he soon began to look more closely at the process of diversification and to ask how it is produced by the struggle for existence. From the following note, written in January 1855, it appears that he was not fully satisfied with his first assumption:

On Theory of Descent, a *divergence* is implied & I think diversity of structure supporting more life is thus implied . . . I have been led to this by looking at a heath thickly clothed by heath, & a fertile meadow, both crowded, yet one cannot doubt more life supported in second than in first; & hence (in part) more animals are supported. This is not final cause, but mere results from struggle, (I must think out this last proposition).[46]

Darwin's resolution of how diversity results from struggle is recorded in a note dated September 23, 1856:

> The advantage in each group becoming as different as possible, may be compared to the fact that by division of ~~land~~ [*sic;* Darwin's cancel] labour most people can be supported in each country – Not only do the individuals of each group strive one against the other, but each group itself with all its members, some more numerous, some less, are struggling against all other groups, as indeed follows from each individual struggling.[47]

In *Natural Selection* and the *Origin of Species* Darwin retained the division of labor as a leading theme in his exposition of divergence. But in his account of the process he shifted the emphasis from the group to the individual or variety within the group. He argued that it follows from the principle that the most diverse can best succeed, that among the varying offspring of a species those that are most different from the others and from the parent form have the best chance of surviving. In this way the advantage of diversity, a group characteristic of the larger group, was translated into the advantage of being different, which is a characteristic of individuals or varieties and hence selectable.[48]

As far as I know, the note of September 1856, just quoted, is the earliest instance of Darwin's drawing an analogy between the processes of divergence and division of labor.[49] In this case the analogy is with the political economists' application of the concept to human society. In *Natural Selection* and the *Origin*, it is with Milne Edwards's division of physiological labor.[50] Since the principle of divergence was in process of formation from 1854 on (one might even say from 1847), it seems of doubtful utility to argue that either usage of the concept was the spark that suggested his principle to Darwin.[51] But the idea of the division of labor did contribute to a subtle but significant shift in Darwin's thinking. In the note of September 1856, as in his published discussion of division of labor, the implication is no longer simply that diversity gives an advantage to wide-ranging species. It is rather that diversification tends always to occur, even in groups inhabiting single regions. It does so by a process of "division of labour," enforced by the struggle for existence, which increases the

Two Darwin notes. Top: November 1854, the problem posed by classification. Bottom: September 1856, a key part of Darwin's solution – the principle of divergence. (DAR 205.5.)

complexity of the natural economy of each region. It is this doctrine that Darwin identified in March 1857 as his "principle of divergence."[52]

Even before this final integration of divergence into the theory of natural selection, Darwin was satisfied that with his principle and his calculations on large genera he had succeeded in explaining the branching arrangement of organisms. "All classification follows from more distinct forms being supported on same area," he wrote in August 1855. The following May, in a note already quoted in part, he said, "How according to descent can be explained this curious arrangement of all living & extinct beings. If varieties differ only from species in amount of difference & less permanence, then let us look in what kind of groups we find varying species. We find them in larger genera."[53] In the *Origin* Darwin presented his explanation of classification as follows: The "dominant species" belonging to the larger genera vary most, thus giving rise to new varieties and species that share the advantages which made the parent form dominant. "Consequently the groups which are now large, and which generally include many dominant species, tend to go on increasing indefinitely in size." But because the offspring of each species try to occupy as many diverse places in the economy of nature as possible, and when they succeed they cause the extinction of the less divergent forms, "the inevitable result is that the modified descendants proceeding from one progenitor become broken up into groups subordinate to groups."[54] It is the inevitability of divergence that makes this a satisfactory solution to Darwin's problem. By showing that division into groups and subgroups is the inevitable result of evolution by natural selection, Darwin brought his theory into harmony with the branching conception of contemporary naturalists, who considered divergence to be a general law of the organic world.

With his new principle Darwin adopted a new theory of the basis of classification. He no longer thought that the existence of groups was due solely to extinction, which created gaps in a potentially continuous "series of connection." Now he said that descent itself, by reason of its necessarily diverging character, creates gaps. In his note of May 11, 1856, he said, "It is often said [as by himself] that all groups are due to extinction, but perhaps not quite correct: if all that had ever lived were put before us, yet there w^d be projecting masses, for in certain lines the birds & mammals c^d not stand. It is owing to descent & extinction."[55]

The keystone of the book

In May of 1856 Darwin began writing *Natural Selection*. Between December 1856 and March 1857 he drafted Chapter IV ("Variation under Nature"), Chapter V ("The Struggle for Existence"), and Chapter VI ("On Natural Selection"). A little over a year later, just before he was interrupted by Wallace's letter, he added a long discussion on large genera to Chapter IV and over forty pages on divergence to Chapter VI.[56] From the portion of Darwin's manuscript that was completed before these additions it is clear that the material they contain was originally destined for later chapters in the book. In the original draft of Chapter VI, Darwin's discussion of divergence occupied only page 27 and the top two or three lines of page 28. Page 27 has not been found, but two existing notes, together with page 28, make possible an approximate reconstruction of its contents. The clearest of the notes reads as follows:

> Every single organism may be said to try its utmost to increase (geometrically), therefore there is strongest possible power tending to make each site to support as much life as possible – How to measure life   Chemical action – How can most life be supported? By diversity ((utilizing different food – like division of labour in organs)) – Explain chemicity [?] facts – isl^d – coral islets – square yard of turf – Better still grasses – Wheats – Heaths – Clovers at Lands End. Results – habit   ultimately structure – Ch. 6[57]

The other, apparently earlier note, whose content is similar, makes it possible to identify this as a sketch of the argument of the missing page 27. The note was originally headed "Divergence Theory (Perhaps only just allude to)," but the parenthetical remark has been lined out and the annotation "alluded to in Chapt 6 p. 27" added.[58] The conclusion of the brief discussion in Chapter VI, at the top of the original page 28, reads, "This principle of divergence, I believe, plays an important part in the affinities or classification of all organic beings."[59]

Darwin's purpose in introducing the principle of divergence in the original Chapter VI appears to have been simply to indicate its relation to the theory of natural selection. The full discussion of the principle was to be reserved for the chapter on classification.[60]

There was therefore no reason to include the discussion of large genera in the early chapters of the book. All that was necessary was that the data be presented before or as part of the explanation of classification. In the original draft of Chapter IV there is no mention of the proposition that varieties occur most often in large genera. In Chapter V Darwin indicated that this would be treated at some later point, and there is a similar reference in Chapter VI.[61] Darwin may initially have intended the discussion of large genera for the chapter on classification itself, but by April of 1857 he had decided to make it the last section of Chapter VII ("Laws of Variation").[62] He could then refer to it when he came to the principle of divergence in his discussion of classification. The calculations on large genera were never added to Chapter VII, however, nor was the full exposition of the principle of divergence saved for the chapter on classification. In the summer of 1857 both acquired a wholly new significance for Darwin, which led him to incorporate them into the very heart of Part I of his book.

In late June or early July Darwin began making additional calculations on varieties in large genera in preparation for his discussion in Chapter VII, which he was then drafting. On July 13 his young neighbor John Lubbock pointed out a fundamental procedural error that vitiated all his statistics. His calculations, including all those made in the fall of 1855, did not prove as he had supposed that there are proportionally more varieties in large than in small genera.[63] Darwin was forced to do them all over, and accordingly, on July 14, he requested again from Hooker all the books he had previously gone through. It was while working out the calculations a second time that Darwin realized that they could be put to a new and much more important use than he had previously thought. His object had been to show that large genera are usually increasing in size. He now concluded that the same data showed that the production of varieties leads to the production of species. In the addition to Chapter IV he explained the argument as follows: If species are separately created, each with a greater or lesser tendency to vary, there is no reason to suppose that proportionally more varieties should occur in a group with many species than in one with few. But if species are formed by variation and natural selection, there is such a reason. In that case a large genus is one in which there has been much variation and the formation of many species. It is one in which circumstances have been favorable for variation, and so we might expect that circumstances would still be favorable and therefore that we would at present find more varieties in it than in smaller genera. "Where many large trees grow, we expect to find saplings," Darwin said. And if we find that in fact there are proportionally more varieties in large genera than in small, we should conclude that this is the cause of their being larger; we should conclude that varieties are incipient species. Darwin illustrated his argument by comparing genera to clans in a nation in which a census is occasionally taken:

> If a nation consisted of clans of very unequal sizes, & if we . . . divided the population into two nearly equal halves, all the large clans on one side, & the many small clans on the other side; we should expect to find, on taking a census at a moderately long interval that the rate of births over deaths was greater in the larger clans than in the smaller; and we should expect to find it so, notwithstanding that we knew that some of the small clans were now rapidly increasing in size & some of the larger clans declining. If we found this to be the case in several nations composed of clans, we should conclude that the greater rate of births over deaths was the cause of the size of the larger clans: & not, for instance, the recent immigration of the large clans. What the rate of births over deaths is to our clans, I suppose the production of varieties to be to the number of species in a genus . . . I may add that if we found any trace of the breaking up of the larger clans into smaller clans, we should infer that this was the origin of any new clans, which, had arisen since ancient historical times.[64]

With the formulation of this argument in 1857 Darwin adopted a new way of stating the fundamental proposition of his theory, that species are made by nature, not directly by God: varieties are incipient species. Although Darwin from the notebook period on had supposed that the varying offspring of one species may form a well-marked variety and finally a new species, he had not previously spoken of varieties as "incipient species." There was a point in doing so now, because he now conceived that his calculations on varieties in large genera gave him a powerful new argument for his theory if the theory were stated as the simple proposition that varieties tend to become species. With the adoption of this centrally important new argument, the principle of divergence also acquired a major new function in Darwin's exposition of his theory. Darwin had previously reached an understanding of how divergence must result from the struggle for existence, which is all that was required for his explanation of classification. But he had not considered the reverse relationship, that the principle of divergence might influence the process of selection. Now he concluded that it must do so. It is divergence that converts incipient species into good and distinct species. How, Darwin asked, is the slight difference that separates two varieties of a species augmented into the greater difference that separates two species, "which must on our theory be continually occurring in nature, if varieties are converted into good species?" The answer is the principle of divergence, the continued selection of "those varieties, which diverge most in all sorts of respects from their parent type . . . so as to fill as many, as new, & as widely different places in the economy of nature, as possible."[65]

The change in Darwin's thinking is well illustrated by the contrast between this and the explanation he gave in the original draft of Chapter IV for the greater difference between species than between varieties. There he said:

> The greater amount of difference between the two species than between the two varieties, may be looked at as simply the result of a greater amount of variation; the intermediate varieties between the two species or between them & a common parent having become extinct. Hence as a general rule, species may be looked at as the result of variation at a former period; & varieties, as the result of contemporaneous variations.[66]

There is no notion here of augmenting differences, except by the passage of time. The main point of the original Chapter IV is that there is much variability in nature, in support of which claim Darwin argued that there is no sharp distinction between individual differences, marked varieties, and species. Species differ from varieties only in having varied less recently. In the additions to Chapters IV and VI, on the other hand, the main point is that varieties are incipient species and that it is the natural selection of the most divergent forms that converts them into distinct species. The principle of divergence has here assumed a role scarcely less important than natural selection itself. The principle of divergence, Darwin said, "regulates the natural Selection of variations."[67]

Darwin's new interest in the principle of divergence and in the data on large genera immediately led him into a further and more ambitious round of calculations, involving the tabulation of many more floras than he had previously used. Two or three weeks after Lubbock showed him the error in his calculations, he wrote Hooker asking him to suggest suitable catalogues. "I am got extremely interested in tabulating, according to mere size of genera, the species having any varieties . . . : the result (as far as I have yet gone) seems to me one of the most important arguments I have yet met with, that varieties are only small species – or species only strongly marked varieties. The subject is in many ways so important for me." On August 22 he wrote Hooker again, this time asking to borrow several floras and again saying how important the work was for him: together with "what I call a principle of divergence" it explains all classification. When on September 5 Darwin sent Asa Gray the letter which in the following July formed part of the Darwin – Wallace presentation to the Linnean Society, divergence seemed to him so important as to require an entire section (one of six) of his abstract of his theory. The emphasis was still on its utility in explaining classification, but Darwin introduced his discussion by saying that the principle of divergence plays "an important part in the origin of species."[68]

The calculations and the borrowing of more and more books continued through the fall of 1857 and the early months of 1858, punctuated by discussions with Hooker and efforts to answer his objections: February 9, 1858: "I am satisfied that there must be truth in the rule that the small genera vary less than the large. *What do you think?*" March 10, 1858: "I am in better heart about my tables of vars. in large & small genera, since trying them in various ways to test your serious objection." March 11, 1858: "The work has been turning out badly for me this morning & I am sick at heart & oh my God how I do hate species & varieties."[69] The project culminated in March and April in the addendum to Chapter IV, which was sent to Hooker for comments on May 6. In April and May Darwin wrote the section on divergence for Chapter VI. The increased importance of the principle of divergence is summed up in a letter Darwin wrote Hooker on June 8, after Hooker had given a favorable initial response to the addendum to Chapter IV:

> The discussion [of varieties in large genera] comes in at the end of the long chapter on variation in a state of nature, so that I have

discussed, as far as I am able, what to call varieties. I will try to leave out all allusion to genera coming in and out in this part, till when I discuss the "Principle of Divergence," which, with "Natural Selection," is the keystone of my book; and I have very great confidence it is sound.[70]

Conclusion

Darwin's remarks on divergence in his *Autobiography* are vague and unhelpful on several matters, such as the date of his solution to the problem, and treacherously ambiguous on the one point on which they seem to be most clear. Although he suggested that the principle of divergence had something to do with classification, one would scarcely suspect from his account that it originally belonged to Part II of his book or that its origin is to be found in a problem that arose in the course of his efforts to translate the theories of his contemporaries into evolutionary terms. Darwin said he had long neglected the problem of divergence, "the tendency in organic beings descended from the same stock to diverge in character as they become modified."[71] The location of his solution in the *Origin* and current debates among evolutionary biologists have made it easy to assume that the problem had to do chiefly with the process of speciation.[72] Darwin in the 1850s did reach a new conclusion about speciation, but this was an incidental consequence of his efforts to explain classification. In order to show that divergence is a general tendency, Darwin modified his earlier views on speciation because they depended on the accidents of transport and geographical isolation, and this made divergence a matter of mere chance. To explain why there is a "tendency" to diverge, Darwin first introduced into his old model of speciation the idea that the most diverse forms can best succeed, which he thought would mean that the varying offspring of forms that spread into new regions would have a better than average chance of giving rise to new species. In the process of integrating this principle into the theory of natural selection, Darwin came to see it as implying a tendency toward ecological diversification and hence ecological rather than geographical speciation.

Darwin's views on speciation and classification are closely connected, but there are good reasons for not conflating them. Had the problem of divergence proved on examination of Darwin's notes to have been the best way to explain the multiplication of species, this might have helped perpetuate the idea that Darwin was an isolated scientist whose aim was merely to construct the best theory to account for the facts as he saw them and that his problems were problems concerning his mechanism for producing evolutionary change, such as how to avoid the swamping effect of blending inheritance. But since the problem was classification, and since it arose as a result of recent innovations in theoretical natural history, the development of the principle of divergence appears as an instance in which the direction of Darwin's speculations was determined by the theories and perspectives of his fellow naturalists. It is an instance of considerable importance, for not only did Darwin see the principle of divergence as a new and better way of explaining the manufacturing of species,[73] but also it played a key role in the development of his theory. Although it originated in Part II of Darwin's species work, the principle of divergence was largely responsible for the transformation of the theory of natural selection that occurred between 1844 and 1859.

8

The principle of divergence and the transformation of Darwin's theory

> When new form introduced into island, it must not be perfectly adapted, else [it] will not vary.[1]
>
> Organic beings seem to be perfect only in that degree required by our theory, namely to be enabled to struggle with all competitors in their native country.[2]

In the "Essay of 1844" the structure of Darwin's theory was largely determined by a network of assumptions that are present in his earliest notebook on transmutation: that in general species are perfectly adapted for their conditions of life; that perfectly adapted forms do not vary; that organic change occurs only in response to environmental change. These assumptions exerted their influence chiefly through Darwin's theory of variation, which effectively confined the working of natural selection to those times and places in which organisms have been subjected to new external conditions. Darwin in 1844 assumed that there is little variation in nature because variation is ultimately dependent on geological change, which is slow and intermittent. Only when geographical or climatic conditions are changing or when organisms are transported to new regions to which they are not perfectly adapted do true species become varying species. In such circumstances natural selection comes into play to produce new, perfectly adapted, and so nonvarying, species. In form, Darwin's theory was still a theory to explain how perfect adaptation is maintained in a changing world. By the time he wrote the *Origin of Species* he had altered his theory of variation in such a way that perfect adaptation ceased to regulate variability. He had concluded that there is much variation in nature and that no change in external conditions is necessary to produce variation, thereby freeing variation and selection from the constraints that had previously limited their operation. By 1859 Darwin had also explicitly rejected the assumption of perfect adaptation in favor of the view that organisms are only well enough adapted to struggle successfully with their competitors.

The several elements that combined to produce Darwin's revised theory had their roots in his speculations of the 1830s and 1840s, but the new theoretical structure only emerged in the course of his work on divergence. The same notes that enable us to trace the development of the principle of divergence also make it possible to determine to within a year or two when, between 1844 and 1859, this shift from old structure to new occurred. When Darwin in November 1854 introduced into his explanation of classification the principle that the most diverse forms can best succeed, he pictured the multiplication of species as occurring under the same conditions as he had envisioned in 1844: A successful species spreads into many regions and is modified in different ways in each of them. Several assumptions about the process of transmutation are involved here, the most important of them being that geographical isolation is generally necessary for the production of new species and that new conditions are required to cause variation.[3] By March of 1857, when Darwin included a page on divergence in Chapter VI of *Natural Selection,* he was working on the assumption that variation and the multiplication of species may occur without geographical isolation or any change in external conditions. This is implicit already in the note of September 1856 in which a group of organisms is compared to inhabitants of a single country who are enabled to increase in numbers by division of labor.[4] It is made explicit in the note which I suggested was a preliminary sketch for the page on divergence in Chapter VI. There the pressure of population is said to be a power tending to make "each site" support as much life as possible. This enforces diversification ("utilizing different food" is mentioned in illustration) entailing changes in habit and "ultimately structure."[5]

If pressed to locate more precisely the point between November 1854 and March 1857 at which the transformation occurred, I would say the second half of 1856. It was the result of Darwin's changing views on several questions, such as what causes variation, how much variation there is in nature, whether important parts vary, how new "places" are created in the economy of nature, and whether geographical isolation is necessary for the multiplication of species. On all of these questions Darwin had for some years been moving toward responses that differed from those he offered in the "Essay of 1844". On all of them his work on divergence persuaded him, partly on empirical, partly on theoretical grounds, that the new responses were preferable to the old; and his successful integration of the principle of divergence into the theory of natural selection in about September 1856 made possible, and in some cases required, the adoption of these new responses.

Geographical isolation and the interrelations of organisms

Early in his speculations Darwin concluded that geographical isolation was probably necessary for the multiplication of species because otherwise crossing would obliterate any differences that might arise.[6] After he decided that variations are "accidental" and that only some are in the right direction to adapt the species to changing conditions, he concluded for the same reason that transmutation can most easily occur in a small population. Where there are many individuals of the species, the few favorable variations will be lost through interbreeding with other variants and with the parent form.[7] Darwin supposed therefore that isolation of a small part of a population from the parent stock (as on an island) would be favorable not only for the multiplication of species, but for the production of any new species. This belief that fewness of individuals favored transmutation was called into question, however, by Darwin's solution in 1844 to the problem of how genera and families arise.

> The diffusion of a species into six regions (or its preexistence in one great region before its division) shows it has some advantage over antagonist species; & yet such diffusion is thus contingent [on] six species of a genus being formed . . . Hence flourishing species must be the parent of new forms & not dying species. But flourishing species generally in great numbers, which is opposed to my view.

Darwin temporarily resolved the conflict by adding, "Flourishing in small district, most fertile source of new forms."[8] But as flourishing species became more and more important for his explanation of classification, he reversed himself on the advantages of a small population and expressed doubts as to whether islands have been very important in the "manufacturing of species." Fewness of individuals retards selection, he said, because there are fewer chances for favorable variations to occur. For this reason it is probable that large land areas have been most important in the production of new species. If a continent breaks up into several large islands, conditions are ideal for variation and natural selection, but such division of the land is not absolutely necessary because there are other means of preventing inter-breeding besides geographical separation.[9]

Between 1844 and 1857 other considerations also were weakening Darwin's belief in the necessity for geographical isolation. Hooker pointed out to him that a single island sometimes has several good species of the same genus. Darwin agreed with him that this was "hostile to descent"[10] and then devoted a page of notes to possible explanations of this anomaly. One of these is especially interesting: "Several species of same genus w^d be apt to arise on same isl^d in proportion as that isl^d was badly placed for new colonists – for then they often w^d have to fill separate functions."[11] This closely resembles the principle of divergence. But for the time being it remained only a special explanation of an unexpected fact. Darwin supposed that the multiplication of species by means of functional (or ecological) isolation might occur in exceptional circumstances, but he was not yet prepared to adopt it as a workable alternative to geographical isolation. This was still the case when he composed the note "Theoretical Geograph. Distrib." in November 1854. When Darwin made his calculations in Schoenherr's catalogue to prove that large genera range widely, he found also that small genera are usually "local" (that is, all their species are confined to a single great region of the earth). Although he must have expected this in one sense – he was looking for a difference between large and small genera in this respect – the results surprised him: "No doubt here comes in question of how far isolation is necessary, & I sh^d have thought more necessary than facts seem to show it is." He reflected that there can never be complete isolation, since parent forms will be present, and he went on to suggest other ways to prevent the swamping of favorable variations. There must be vigorous selection of "some isolation, from habit, fewness [of individuals, or] nature of country," he said.[12]

Over the next two years the idea of "isolation from habit" became increasingly attractive to Darwin. His work on divergence led him to suppose that if diversity is an advantage, then population pressure will act in every group as a powerful impetus forcing its members to diversify so as to be able to fill as many different places as possible in the economy of the region they inhabit.[13] Darwin could argue that diversification may occur in a group inhabiting a single region because he was willing to consider nongeographical modes of isolation. At the same time, his principle of divergence reinforced his feeling that the adoption of new habits might be an effective bar against interbreeding with the parent stock. In the principle of divergence Darwin gave expression to a view of the economy of nature in which the interrelations of organisms take precedence over geographical and climatic considerations. He urged that the "places" organisms may occupy are defined less by external features and more by the habits (broadly conceived) of organisms, and this made much more plausible the idea that physical barriers are not necessary to produce reproductive isolation. Whereas in 1854 Darwin had been surprised at the results of his calculations in Schoenherr, after he developed the principle of divergence he took pains to insist that "it must not . . . be supposed that isolation is at all necessary for the production of new forms."[14]

Darwin's belief in the primacy of the interrelations among organisms in defining "places" was new in 1856. It is not that he had ignored their role previously. On the contrary, he had consistently emphasized that the ranges of species are limited not by climate, but by other species; that the relations of interdependency among organisms are the primary constituents of the complex of conditions in which the struggle for existence occurs; and that the introduction of a single new species may substantially alter the economy of a region ("Larch plantation at Maer best case of one species altering the whole adaptation of other species").[15] But Darwin nevertheless continued to be bound by the common assumption that species are most importantly related to the physical conditions of life, that "temperature [is the] greatest ruling cause of differences in organisms."[16] Not until 1856 did changes in associated organisms surpass in importance physical conditions of life in Darwin's ranking of factors favorable for the operation of natural selection.

In the fourth transmutation notebook, in a discussion of the causes of organic progress, there is a remarkably full anticipation of the conception of organic relations that subsequently appeared in Darwin's exposition of the principle of divergence: "The enormous *number* of animals in the world depends [on] their varied structure & complexity. . . . hence as the forms became complicated, they opened *fresh* means of adding to their complexity . . . if we begin with the simplest forms & suppose them to have changed, their very changes tend to give rise to others."[17] The picture here presented is of organic change stimulating further organic change. Organisms are adapted principally to each other, and so there is no need for changes in external conditions to initiate changes in organic structure or to produce new situations to which organisms must adapt. This vision of nature does not reappear in Darwin's writings until the mid-1850s. It was neglected for eighteen years because it was wholly at odds with the view that Darwin shared with most of his contemporaries, that organic change – whether it be new creations, the production of varieties, or the production of new species by transmutation – occurs primarily as a response to changes in physical conditions: Darwin continued to see organic change from the perspective of natural theology, as a means by which the organic world remains in harmony with the ever-changing physical world. It conflicted also with the closely related proposition, which Darwin derived from Lyell, that the number of species in the world is approximately constant over time. Lyell's nondirectional theory of the earth, combined with his traditional perspective on adaptation, produced a nondirectional conception of the history of life. Darwin adopted this Lyellian view of constancy (though he rejected Lyell's nonprogressionism) while still on board the *Beagle*. In March 1839 he drew on it to scuttle his own suggestion, just quoted, that organic complexity and population may increase indefinitely: "In early stages of transmutations, the relations of animals & plants to each other would rapidly increase, & hence number of forms, once formed, would remain stationary, hence all present types are ancient."[18] After the dawn of life the world will have filled up rapidly, and from then on the number of species and places will have been constant.

Darwin long remained wedded to Lyell's view of the "fitness" of things. Although he defined the situation occupied by an organism by both physical and organic conditions, he continued to treat organic conditions and their changes as subordinate to physical conditions and their changes.[19] He assumed that the number of places is constant, that the birth of one species requires the death of another. Even in his favorite example, of a form's being introduced into a rising island, this attitude prevails. The intruder will modify the conditions of the previous inhabitants, and new species will be selected for new ends. But there is no hint that the complexity of the island economy will be increased. The potential places in a region and in the whole world are fixed and predetermined by geological conditions. As a result of the introduction of the new form into the island, all or most of the inhabitants will be more strictly limited to some of these potential places, and the new form and its descendants will fill one of the potential places that was previously unoccupied or imperfectly filled. But organic changes will not themselves create new places. For all his insistence on the complexity of the organic conditions under which the struggle for existence goes on, Darwin's conception of place in 1844 was still essentially geological. While new places are being formed and filled on the rising island, he said, other places elsewhere are simultaneously being destroyed ("when one part of the earth's crust is raised it is probably the general rule that another part sinks"). Where Darwin in 1844 spoke of adaptation to conditions, he was nearly always referring to physical conditions. When he mentioned changes in conditions that caused species to be less than perfectly adapted, again he usually meant changes in physical conditions. And the places on a rising land that he supposed new species would occupy were most often the "stations" of contemporary biogeography, which were defined in geological and geographical terms.[20] In short, Darwin's concept of place, which in *Natural Selection* and the *Origin* closely resembles, as others have noted, the modern concept of niche, was in 1844 not very clearly articulated; its primary reference was to physical, not organic, conditions; and it reflected traditional attitudes about the dependence of organic on geological and climatic change.[21] This is what Darwin was referring to when he wrote Lyell in 1859 that it had taken him many years to overcome the habit of attributing too much influence to climate. The importance of the interrelations of organisms "you will perhaps think very obvious," he said; "but, until I repeated it to myself thousands of times, I took, as I believe, a wholly wrong view of the whole economy of nature."[22]

In November 1854, as in the "Essay of 1844," Darwin still treated interrelations of organisms as subordinate in importance to external conditions.[23] But in the summer of 1856 he introduced into his chapter on geographical distribution the proposition that "forms of being" are "infinitely more related to each other" than to physical conditions.[24] The fact that two months later Darwin was comparing divergence to the division of labor in a single country (that is, without, necessarily, any geographical isolation) indicates that it was at about this time that he arrived at a new conception of place and a new estimation of the relative importance of organic and physical change. It was probably during these months in 1856 that he made the following addition to a note on divergence originally written in January 1855: "Always remember how organisms are most importantly related to other organisms . . . The number of species . . . goes on increasing in geometrical ratio – because relations get more & more complicated."[25] This is but a brief expression of the vision of nature we find in *Natural Selection:*

> The structure of each organism stands in the most direct & important relation to many other organic beings, and as these latter increase in number & diversity of organisation, the conditions of the one will tend to become more & more complex, & its descendants might well profit by a further division of labor.[26]
>
> . . .
>
> I infer that the association of an organism with a new set of beings, or with different proportional numbers of the old inhabitants, [is] perhaps the most important of all elements of structural change. If a carnivorous or herbivorous animal is to be modified, it will almost certainly be modified in relation to its prey or food, or in relation to the enemies it has to escape from. Change of climate will act indirectly in a far more important manner than directly, namely in exterminating some of the old inhabitants or in favouring the increase of others. The immigration of a few new forms, or even of a single one, may well cause an entire revolution in the relations of a multitude of the old occupants. If a certain number of forms are modified through natural selection, this alone will almost certainly lead to the modification of some of the other inhabitants. Every where we see organic action & reaction. All nature is bound together by an inextricable web of relations; if some forms become changed & make progress, those which are not modified or may be said to lag behind, will sooner or later perish. – [27]

In this new conception of nature as a system of "organic action & reaction," inorganic conditions are still supposed to set ultimate limits to diversification ("inorganic conditions . . . do not tend to become infinitely more varied") and to the amount of life that can be supported. But until those limits are reached – and Darwin did not believe they have been – inorganic concitions do not keep the number of places or of species constant, nor are changes of physical conditions the primary stimulus to organic change.[28]

Despite the novelty of his conception of the struggle for existence, Darwin's assumption in 1844 that inorganic conditions are the most important determinants of organic structure was very similar to the view of most of his contemporaries (on whichever side of the debate over teleological explanation they stood, naturalists tended to think of adaptation as most importantly related to climate). Consequently "place" was still for him, as for them, a rigid, static concept. "Adaptation" was similarly rigid, the norm in nature being perfect adaptation to place. In *Natural Selection*, on the other hand, adaptation has reference chiefly to other organisms. Adaptation is not rigid or absolute, but plastic, flexible. It depends on the nature of places, and places are plastic. They change their shapes and increase in number with every progressive increase in the division of labor.[29] It is instructive in this regard to contrast Darwin's enumeration in the "Essay of 1844" and in *Natural Selection* of the circumstances favorable for selection. In 1844 the existence of unfilled places is not on the list. The reason is that Darwin's theory was then still a theory of response to external change. The most important factor he noted was a change in conditions, which he considered to be necessary to produce variation.[30] Places were of no special concern because Darwin assumed that these same changes in conditions will open places – usually by causing extinction. In *Natural Selection*, the existence of places is first on the list.[31] Place is all-important because Darwin no longer assumed that the process of variation is initiated by changes in conditions that will also automatically open new places. Organic change may occur independently of geological change. Therefore what determines whether natural selection can produce a new form is whether places have been opened, either by the association of new organisms; by modification of other forms, causing an increase in the complexity of organic relations in a region; or by some other means.

Camille Limoges has insisted on the importance of Darwin's

"ecological conception" of nature and has noted also its association with the principle of divergence. But contrary to Limoges's suggestion, this new vision of nature was not implicit in Darwin's early speculations on geographical distribution or in the theory of natural selection.[32] For over eighteen years Darwin's perspective on place and adaptation was similar to that of his contemporaries, and I see no reason to think that it would not have remained so had Darwin not developed his principle of divergence.

Variation

The one factor that Darwin recognized in *Natural Selection* as more important than the existence of places for a new form to occupy was the occurrence of variation, "the basis on which the power of selection rests."[33] By this time variability was no longer a concern for Darwin. In 1844, however, it had presented a major problem. He believed then that perfectly adapted forms do not vary. He believed in consequence that species, which are normally perfectly adapted, normally do not vary and that there is therefore little variation in organisms in a state of nature. And he considered changes in external conditions to be a prerequisite for evolutionary change because without them there would be no variations on which selection could work. As a result, Darwin's discussion of variation in nature in the "Essay of 1844" is dominated not by evidence of variation, but by an argument – that if he is right about the causes of variation under domestication, then there must be variation in nature, even though we see little of it. By the time he wrote the corresponding chapter of *Natural Selection* in the winter of 1856–7, Darwin's outlook had changed. The bulk of the chapter is devoted to proving, by listing instances, that there is much variation in nature. By way of introduction Darwin presented what was for him a new view of the causes of variation, which said in effect that variation may occur at any time, with or without changes in external conditions. It was this new view that made possible Darwin's conception of the evolutionary process as "organic action & reaction." It made natural selection virtually independent of external changes and eliminated what I previously called the natural theological structure of Darwin's 1844 theory.

It is tempting to attribute the change in Darwin's theory of variation simply to the evidence he was able to collect and particularly to his first-hand acquaintance with barnacles. Undoubtedly this was important. One cannot ignore his statement to Hooker on the subject, with its direct rebuttal of the opinion he had held in 1844:

> You ask what effect studying species has had on my variation theories: I do not think much – I have felt some difficulties more. On the other hand, I have been struck (and probably unfairly from the class) with the variability of every part in some slight degree of every species . . . I had thought the same parts of the same species more resemble (than they do anyhow in Cirripedia) objects cast in the same mould.[34]

But there are reasons to think that this new fact of the variability of cirripedes was not alone sufficient to cause the radical change that took place in Darwin's view not only of the amount, but of the cause, of variability (this is partly because differences in the surfaces to which they are attached are a constant source of change in the external conditions of life of each species).[35] Three years later, in 1852, Darwin was still very cautious, even though by this time his reading, as well as his barnacle work, had supplied him with a considerable store of facts on variation: *"When I see that species even in a state of nature do vary a little,"* he said, *"& seeing how much they vary when domesticated, I look with astonishment at a species which has existed since one of the earlier Tertiary periods . . . This fixity of character is marvellous. – Identity is the law, & a great effort is required to produce any change."*[36] Darwin was beginning to think he could show that there is some variation in nature, but clearly he still believed it was uncommon, and clearly his theory of variation was still the same as in the "Sketch" and "Essay." It remained unchanged in 1854. It was the basis for Darwin's speculations on divergence in November of that year;[37] and it is also the theory he alluded to in his volume on sessile cirripedes (published late in 1854).[38]

Although the cirripede work did not immediately produce any great change in Darwin's views on the prevalence or causes of variation,[39] it was almost certainly one of the factors that gave him the confidence by the end of 1856 to assert that "there is much variability in organic beings in a state of nature."[40] This meant that he did not have to rest his case on the weak argument from the supposed causes of variation in domestic productions. In Chapter IV of *Natural Selection* the discussion of the causes of variation is of minimal importance. The emphasis instead is on the evidence. By this time, however, there were other factors at work too, including a new theory of the causes of variation which, in contrast to his old theory, would lead one to expect to find much variation in nature. The impetus behind Darwin's new theory of variation was not his cirripede work (which might be said rather to have reinforced his theory) but rather the work that went into his development of the principle of divergence. In two respects this undermined his long-held assumption that variation is a response to geological and climatic change. First, from 1847 on, and more so from 1855, Darwin's investigations of variation in large genera suggested that variation is not dependent on external change, that it is rather a property especially characteristic of large genera, whatever the conditions under which their species live.[41] Second, Darwin's elaboration in the period 1854–6 on the idea of diversification of a group in a single region suggested the possibility of organic change under nearly uniform conditions. The full development of this idea, culminating in the proposition that divergence occurs in a "country not undergoing any change,"[42] demanded a new theory of variation, in which altered external conditions are not held to be necessary for its production. Such a theory is presented in *Natural Selection*.

Darwin's discussions of the causes of variation in the "Essay of 1844" and in *Natural Selection* are similar in many respects, but a close comparison reveals two significant differences. One, a mere shift in emphasis, is crucial. At a single stroke it freed natural selection from dependence on external change. Where in 1844 Darwin assumed that there is no variation without a change in conditions, in 1856 he assumed that there will be variation unless conditions remain absolutely constant.[43] In this formulation variation is in effect, though Darwin would not admit it in principle, an innate property of organisms.[44] Variation may be produced by any of the slight alterations in circumstances – a colder winter, a drier summer, new associated organisms – that are constantly affecting all species. "Everywhere," he said, "organic beings present individual differences, & some few more marked variations . . . there is no land, so well stocked with organic beings, or with conditions so unvarying, but that in the course of time, natural selection might modify some few of the inhabitants & adapt them better to their place in the great scheme of nature."[45]

The second difference, though less significant, shows even more clearly the role of Darwin's work on divergence in altering his views on variation. In the "Essay of 1844" Darwin considered and found inadequate Andrew Knight's hypothesis that an increase of food is a principal cause of variation under domestication. Darwin needed an explanation of variation in domestic productions that would allow him to argue, with little evidence, that there must be variation in nature. He could not easily prove that organisms in a state of nature sometimes have a surplus of food (his application of Malthusian principles might seem to imply the opposite).[46] He could, however, prove that external conditions change, merely by appealing to the well-established facts of geology. Moreover, Darwin's most firmly entrenched assumptions about nature told him that the cause of variation must be environmental change. In the first transmutation notebook he said that the purpose of variation is to adapt organisms to change, and in the 1840s he continued to think that variability is induced by the same changes that cause organisms to be imperfectly adapted. But in the mid-1850s his work on the principle of divergence gave him a good theoretical reason for thinking that no external change is necessary to produce variation – and for adopting Knight's hypothesis: "Again Knight's remark about excess of food causing *variation* – It w^d agree admirably with my views that *flourishing* tribes varied most."[47] In *Natural Selection* an increase of food is mentioned prominently as a cause of variation, particularly in connection with Darwin's views on dominant species.[48] "Change in conditions" was coming to mean for Darwin not primarily a change in physical conditions, but rather any slight change in circumstances, with an increase in food as one of the most important.[49] A decade later, in *The Variation of Animals and Plants under Domestication* (1868), he considered excess of food to be "probably the most powerful" cause of variability. Geological and climatic changes were merely among the many other possible causes. "A change of almost any kind in the conditions of life," Darwin said, "suffices to cause variability – different changes acting on different organisms."[50]

This reassessment of Knight's hypothesis reflects no change in the available evidence on the causes of variation. Darwin simply adopted in 1856 the explanation that best met the requirements of his theory. When his views on flourishing species and division of labor made an increase in food look attractive and a change in external conditions seem unnecessary, he altered his theory accordingly. The ease with which the alteration was made suggests that his choice of explanations was not governed, as it was in 1844, by any residual assumptions about the purposes of variation in a self-adjusting and stable economy of nature.

One other aspect of Darwin's new theory of variation requires comment. In *Natural Selection,* for the first time, the expression "individual differences" has the same meaning that it has in the *Origin* and in Darwin's later writings. I noted earlier that in the transmutation notebooks Darwin supposed that the slight fluctuations in conditions that constantly occur produce adaptive alterations in organisms but that these "individual differences," as he called them, are normally obliterated by crossing, which is a good thing because it facilitates adaptation to the "great changes" in conditions that geology brings about. In 1844 Darwin made a similar distinction. He said that in nature individuals often differ slightly in external and unimportant parts, and he contrasted this with the "more important" variation that occurs after a change in conditions. It was the latter, he thought, that was useful for selection, for only with external change would the organization of the species become plastic, so that even important and internal parts might vary.[51] In *Natural Selection* individual differences are any differences, no matter how slight, in individuals of the same species. They are likely to be produced at any time by subtle changes in conditions, and they may occur in important as well as unimportant parts. There is no fundamental difference between them and well-marked varieties, and when added up in one direction by selection they constitute a variety. They are the foundation for the operation of natural selection.[52]

In Darwin's new theory perfect adaptation is no longer an effective regulator of the production of variation. Darwin's old equation of imperfect adaptation and variability no longer had any meaning for him. How could one say that a group of perfectly adapted animals that ate abnormally well for some time and began to vary had been made imperfectly adapted by the excess of food? Darwin, it is true, refused to say that variation is a necessary consequence of reproduction, insisting that it must have definite – external – causes: "variability *must* supervene & perhaps cease."[53] The fact that he maintained this opinion indicates perhaps the persistence of a part of his old attitude – that the essence of a well-adapted species is stability, that some cause is required to produce change. But with his new views about the prevalence of variation in nature and his new emphasis, that variation is to be expected unless conditions remain absolutely constant, this attitude ceased to be of much importance for the operation of natural selection. There is always enough variation to enable selection to work.[54]

Relative adaptation

Around 1840, while considering the extent to which organisms are adapted to other organisms, Darwin reflected that "the perfection of adaptation in two countries may possibly vary,"[55] In this early note Darwin, it might be said, was groping toward a conclusion which the conceptual tools he was then using made it difficult, if not impossible, for him to reach. He still assumed that natural selection makes every species perfect for the place it occupies and that perfectly adapted forms do not vary. But insofar as they are defined by other organisms, places, he considered, might vary among themselves in the perfection they allow to the organisms that are produced to fill them. Perhaps then one species that is perfectly adapted for its place will be less perfect than another species perfectly adapted for some other (less perfect?) place. Merely to state the proposition in this form is to indicate the difficulties it involves and to suggest why it did not become important for Darwin until many years later. In the "Essay of 1844" there is no mention of the possibility of there being different degrees of adaptedness among organisms in different places,[56] and in 1846 Darwin still held that perfect adaptation governs variability.

At about the same time that perfect adaptation ceased to be a theoretically important assumption in Darwin's thought, he concluded that adaptation is relative. In "Theoretical Geograph. Distrib." (November 1854), in discussing increasing and dying genera, Darwin argued that the inhabitants of two different continents are probably not equally well adapted and that if put into competition with each other the productions of one continent would probably "prevail considerably" over those of the other.[57] During the next three years, with the full development of the principle of divergence, relative adaptation became a necessary implication of his theory. This is true in two different senses. First, the adaptedness of every species is relative to the adaptedness of other species. Because organic conditions are the principal determinants of places in the natural economy, the adaptation of every species depends on the adaptation of those others that together constitute the conditions against which it "struggles" in Darwin's "large and metaphorical sense" for survival. Since it is no longer supposed that variation ceases when a species has been made perfectly adapted for its place, the only limit to the amount of change natural selection can produce is the ability to struggle successfully. Improvements beyond those barely necessary for competing successfully give a species no advantage in the struggle, and therefore the struggle cannot cause such improvements to be selected. At any given moment, then, every species is only "as perfect as, or slightly more perfect than, the other inhabitants of the same country with which it has to struggle for existence."[58] But – and this is the second sense of relative adaptation – all of the competing species that at a given moment are approximately equally well adapted may become still better adapted in comparison with, say, the species of another country. There is always room for improvement. If variation does not cease when species are well adapted relative to their competitors, natural selection can continue to work indefinitely. It can always make forms better adapted by increasing the ecological specialization, or division of labor, in a group. Diversity is a quality that can always be augmented, and since diversity is an advantage, population pressure will tend to produce more of it. And if a group is improved in this fashion, its competitors will tend to be improved also: "As all organic beings are striving, it may be said, to seize on each place in the economy of nature, if any one species does not become modified and improved in a corresponding degree with its competitors, it will soon be exterminated."[59]

The relativity of adaptation is first clearly asserted in portions of *Natural Selection* that were drafted in March 1857. In referring to the success of introduced forms, which he had long used as an argument against the teleologists' view of adaptation, Darwin said, "Each being in its native country no doubt is adapted to its conditions of existence as perfectly as the other coinhabitants . . . but not one country, still less not one island can be named which does not possess many organic beings naturalised thoroughly well as far as we can judge."[60] The fuller statement quoted at the beginning of this chapter was written the following September, when Darwin was feverishly reworking his calculations and reassessing the role of the principle of divergence in his theory.[61] In the section on divergence that he added to Chapter VI in the spring of 1858, Darwin suggested the possibility of indefinite (though not unlimited) improvement by division of labor.[62] Then, having concluded that adaptation is relative, he gradually over the next few years eliminated from his works much of the language of perfect adaptation. For instance, in his comparison of man's selection with nature's in the notebooks, the "Essay of 1844," and *Natural Selection,* he said that natural selection produces "perfect adaptation to the conditions of existence." In the *Origin* this was altered to say that nature's productions are "infinitely better adapted" than man's.[63]

Its late emergence in Darwin's thought indicates that relative adaptation is not, as might be supposed, a necessary implication of the theory of natural selection. As long as one assumes that "place" is defined primarily by inorganic conditions; that changes in inorganic conditions are required to initiate the process of transmutation; and, most importantly, that the process continues until, and only until, new perfectly adapted, nonvarying forms are produced, there is no incompatibility between the theory of natural selection and the concept of perfect adaptation. By the mid-1850s, however, these assumptions no longer seemed necessary to Darwin. In the course of his work on divergence they played a diminishing role in his speculations, until at last all were abandoned.

Conclusion

Susan Cannon has expressed well the nature of the assumptions that gave Darwin's theory in the 1840s its natural theological structure: "Theological ideas were of importance in shaping Charles Darwin's thought, in part because he did not always recognize them as theological ideas, but only as obvious characteristics of the natural world which needed to be explained."[64] Preplanned harmony, which others saw as an obvious characteristic, Darwin rejected at an early period. But assumptions he perhaps did not recognize as theological, such as perfect adaptation, persisted. In the "Essay of 1844" the idea that transmutation serves to adapt organisms to a changing world is nowhere explicitly stated. Indeed, that was no longer the main point of Darwin's theory. But because he assumed that adaptation is perfect his theory retained the basic structure of a mechanism for responding to external change. As Darwin's work on barnacles and on variation in large genera gave him a more flexible view of variation, perfect adaptation became a less and less effective regulator of variability. The old structure of his theory grew progressively weaker until it was finally eliminated at the time of the incorporation of the principle of divergence into his scheme of nature. In the mid-1850s Darwin concluded that there are no such things as perfectly adapted forms, either in the sense of forms that do not vary or in the sense of forms that cannot be improved. Variation, selection, and the production of better adapted forms may occur in any time or place, with or without changes in physical conditions.

This transformation of Darwin's theory was not effected by a sharp departure from previous lines of thought. Even if, as I have argued, there was a turning point in the latter half of 1856, it was the culmination of a long series of developments stretching back at least as far as the mid-1840s; and it gave rise to further changes during the next several years.[65] Rather than a single decisive break we find numerous shifts in emphasis, some obvious, others subtle. In 1844 Darwin supposed that change in the organic world is inherently intermittent; in *Natural Selection* he implied that it is potentially – but probably not actually – continuous. In 1844 he supposed that inorganic conditions are the principal determinants of structure; by 1856 he considered them less important than organic conditions. When he wrote the "Essay of 1844" Darwin assumed that variation occurs only as a result of environmental change; in *Natural Selection* he assumed that it is always occurring because conditions never remain absolutely unchanged. In 1844 an increase in food is allowed to be perhaps a minor cause of variation; in *Natural Selection* it is said to be an important cause; and by 1868 Darwin called it probably the single most powerful cause. In 1844 Darwin considered islands to be the most favorable sites for the production of new species; in *Natural Selection* he said they are in some ways favorable, but all things considered they are less important than large land areas. In the "Essay of 1844" natural selection is supposed to produce perfect adaptation; in *Natural Selection* it produces forms only as perfect as their competitors. In 1844 Darwin saw nature as a system of inorganic change and organic response; in *Natural Selection* he saw everywhere organic action and reaction. Together these shifts in emphasis, and the numerous others that could be added to the list, produced a substantially new conception of the evolutionary process. It is the Darwinian conception we are familiar with from the *Origin of Species*, but it is not a conception that we are justified in reading back into Darwin's earlier writings. It emerged, after 1844, out of Darwin's complex creative response to the thought of the leading naturalists of his day.

9

Natural selection and "natural improvement"

> "Little men do not turn into tall men"
> (but children do   C. D.)
> > Darwin's notes on Hugh Miller's
> > *Old Red Sandstone*[1]

The transformation of Darwin's theory that occurred in the mid-1850s originated in his efforts to explain certain propositions associated with the developmental conception of organic nature that had gained prominence during the previous three or four decades. In working out his explanation Darwin produced a fusion of that developmental conception and his theory of natural selection. In 1844 natural selection was not a theory of organic development, but simply a theory of organic response to environmental change. Although Darwin believed that the history of life has been a history of development, as his earliest notes on embryology and morphology indicate,[2] until the 1850s natural selection provided no satisfactory explanation of this belief. The principle of divergence, however, gave him one, and in so doing made the theory of natural selection into a theory of progressive development as well.[3]

The principle of divergence depended on recently elaborated definitions of specialization – particularly Milne Edwards's division of physiological labor – which were at the same time definitions of degree of development, of the extent to which a form has departed from the archetype. Consequently, the effect of Darwin's adopting the principle of divergence was to introduce into the theory of natural selection a developmental tendency – not a Lamarckian internal organic force, but a tendency for the large majority of lines of descent to show increasing specialization and complexity. Although this gave a substantially new character to Darwin's theory, which was originally only a theory of adaptation, the change was neither unwelcome to him nor entirely unintended. A major problem for Darwin since the period of the transmutation notebooks had been how to show, without calling on a Lamarckian tendency, that transmutation will necessarily produce progress. Before 1857 his speculations had suggested several theories of progress, but none was entirely satisfactory. Once he worked out the principle of divergence, however, he could argue for the inevitability of progress by pointing to the agreement, which was not fortuitous, between the kind of organic change that his principle said was most likely to occur and the new definitions of "highness" offered by Owen, Milne Edwards, von Baer, and others. This allowed him to claim, as he had long wished to do, that natural selection will "inevitably lead to the gradual advancement of the organisation of the greater number of living beings throughout the world."[4]

Darwin's theory of progress

Although Darwin in many respects was a disciple of Lyell, Lyell's nonprogressionism never held any attractions for him.[5] In the earliest pages of his first transmutation notebook Darwin expressed his belief in progress, and he returned again and again to the question of the relationship between transmutation and improvement. "Each species changes. Does it progress. Man gains ideas. The simplest cannot help becoming more complicated; and if we look to first origin, there must be progress . . . Every successive animal is branching upwards different types of organisation improving as Owen says simplest coming in and most perfect and others occasionally dying out."[6] In the fact that the embryo during development usually rises in organization Darwin saw proof of progress in past ages and a guarantee that progress is the rule in organic change. "An originality is given (and power of adaptation is given by *true* generation), through means of every step of progressive increase of organization being imitated in the womb which has been passed through to form that species," he said. "If generation is condensation of change, then animals must tend to improve." "As Larva may be more perfect (as we use the word) than parent, so may species retrograde, but these facts are rare." "There is an analogy between caterpillars with respect to moths, & monkeys & men, – each man passes through its caterpillar state.

The monkey represents this state."[7] Embryology continued to carry the same implications for Darwin after he rejected recapitulation in favor of von Baer's view of development. According to Darwin's explanation of embryonic resemblances, the simple condition of the embryo is a picture, more or less obscured, of the simple condition of the ancestor. "It may be argued with much probability," he said, "that in the earliest and simplest condition of things the parent and embryo must have resembled each other, and that the passage of any animal through embryonic states in its growth is entirely due to subsequent variations affecting *only* the more mature periods of life."[8] "Embryology shows there has been advance."[9]

Darwin never seriously doubted that progress has been the general rule in the history of life. His theory, he said, required it. Animals must have been preceded by vegetables, which Darwin was sure were "low" by comparison. Paleontology showed that reptiles preceded mammals. And Darwin's morphological views told him that if the members of each class have descended from a common ancestor, that ancestor ought to have been less specialized than its various offspring.[10] I do not mean to say that any of these are very good reasons for calling the history of life "progress," "improvement," or "advance." But Darwin thought they were. To him, it was inconceivable that transmutation might be generally degradation. Very simply, Darwin *believed* in progress, and it is not difficult to imagine the web of religious, political, social, and economic attitudes and interests that might have helped produce and sustain this particular view of nature.

Darwin's problem, which he recognized as such by mid-1838, was to explain what causes the progress of organization that embryology, paleontology, and his theory all seemed to say has occurred. Before the mid-1850s there is apparent in his thought, as in Lamarck's, a tension between progress and adaptation.[11] Darwin saw that the natural context in which, as he believed, transmutation takes place excluded the possibility of an innate tendency of organisms to progress regardless of external conditions. He supposed that the impetus for change in the organic world was not the achievement of higher levels of organization, but rather the maintenance of adaptation by the production of new forms suited for new conditions of life. In this his emphasis was the reverse of Lamarck's. For Lamarck adaptation was the cause of deviations from the primary upward movement of life. But from Darwin's perspective, in which adaptation was the central concern, progress was a secondary consequence of adaptive change. There could be no question of an independent *pouvoir de la vie*. Adaptation was always primary for Darwin; but it was only while he believed that adaptation was perfect, and in consequence that transmutation was an intermittent process of organic adjustment, that the tension between adaptation and development was particularly acute. Because a continuously operating developmental tendency in nature was incompatible with his predivergence theory, it was necessary for him, as a believer in progress, to show that an advance of organization will tend to occur as a result of repeated adaptive responses to environmental change.

Although he long considered it a secondary consequence of adaptation, progress was never for Darwin merely an accidental by-product of transmutation. Until some time after he read Malthus he thought progress, including the production of man, was a preplanned outcome of the creator's laws. "Progressive development gives final cause for enormous periods anterior to man," he said. At first he supposed the law of progress was that a species will tend to produce the next higher form in its line of development ("If all men were dead, then monkeys make men. – Man makes angels"). But he soon decided that each species is the result of a particular set of circumstances that is not likely to be repeated. Monkeys and men might both give rise to other species, but probably no particular species, including man, will be produced twice.[12] Once Darwin had concluded that natural selection works on chance variations, he could no longer treat man or any other species as a preplanned, an inevitable, goal of the process of transmutation. But he continued to think that progress in general was inevitable, and he appears to have believed that the production of some intellectual being was inevitable.[13]

After he hit upon natural selection, therefore, as well as before, Darwin was concerned to show that progress was a necessary general consequence of organic change. Between 1838 and 1844 he made numerous attempts to work out an explanation that would do so:

There must be some law that whatever organization an animal has, it tends to multiply & improve on it. – Articulate animals must articulate, & in vertebrate tendency to improve in intellect [August 17, 1838]. Is there some law in nature an animal may acquire organs, but lose them with more difficulty . . . & hence the *improvements* of every type of organization. Such law would explain everything. – *Pure hypothesis* be careful [September 2–6, 1838]. See if any law can be made out, that varieties are generally additions, & not abortive: with reference to the non-necessity of the so-called progressive tendency law [December 1838]. The enormous *number* of animals in the world depends [on] their varied structure & complexity. – hence as the forms became complicated, they opened *fresh* means of adding to their complexity. – but yet there is no *necessary* tendency in the simple animals to become complicated although all perhaps will have done so from the new relations caused by the advancing complexity of others . . . Considering the Kingdom of nature as it now is, it would not be possible to simplify the organization of the different beings . . . without reducing the number of living beings – but there is the strongest possible [tendency?] to increase them, hence the degree of developement is either stationary or more probably increases [January 1839]. I presume, from my theory, as long as any structure can be handed down without being absolutely injurious or requiring nutrition it will be so handed down, as mammae of men, callosities on camels & horses – & therefore probably any structure would rather become accommodated to new circumstances than it would be eliminated, & hence this application of structures to purpose after purpose would tend to render complex the series [probably 1839–42]. From the strong and general hereditary tendency we might expect to find some tendency to progressive complication in the successive production of new organic forms [1844].[14]

Three things are worth noting about this series of speculations. First, Darwin was consistent in his view that there is no innate tendency for a species to give rise to others more highly developed than itself; degradation is always possible, though it is rare. Second, with the exception of the hypothesis on increasing ecological complexity, which Darwin only returned to in the mid-1850s, his speculations centered not on natural selection but on the laws of generation, variation, and heredity. Third, the explanation of progress was a persistent problem for Darwin from 1838 on.

Darwin's concern to account for progress is displayed also in his efforts to find a workable definition of "high" and "low" in organic nature. This was a problem that attracted the attention of post-Cuvierian biologists generally, and it was in the work of his professional colleagues that Darwin found what he needed.

Cuvier's objections to the chain of being and to all developmental conceptions in biology led him to reject in principle (though not in practice) all serial arrangements of animals. Every animal is as perfect as every other, his teleological principles said, in that the organization of each satisfies its peculiar conditions of existence. For those of Cuvier's followers who wished to uphold progression-ism in paleontology, this posed serious problems. In what sense is there progress if all forms are perfectly adapted? Most naturalists in the 1830s, while agreeing that all organisms are perfectly adapted, nevertheless recognized various degrees of complexity of intellectuality among animals.[15] Very often it was admitted that one should only speak of progress within a single *embranchement,* which avoided the question of whether bees or cephalopods were higher (it was clear to all, however, that the vertebrates as a group were higher than either molluscs or articulates and that man was the pinnacle of creation). Sometimes it was said that the fossil record showed a gradual approach toward the present creation, the assumption being that the existence of man and his relatives made the present creation the highest. And there was sometimes said to be a succession of animals with more highly developed nervous systems, culminating in the mammals and man. Clearly Cuvier's strictures did not prevent naturalists' speaking of prog-ress or degrees of perfection. But for those who accepted it, his teleological method precluded the invention of any general criterion of "highness" and "lowness," with the result that for a time there was little agreement as to what progress was. And those definitions that were commonly assumed were of no use to Darwin, for it was not obvious that either mere complexity or a resemblance to man would give an organism any advantage in the struggle for existence.

The naturalists who worked out general definitions of degrees of perfection in the animal kingdom were among those who rejected Cuvier's teleology in favor of a morphological and developmental approach. They too spoke of organisms as perfect-ly adapted for their places in nature, but they said that the various modifications of any one type of structure, such as the vertebrate, might be ranked according to the degree to which they depart from the simple type form. Von Baer defined grade of develop-ment as degree of histological and morphological differentiation. The animal is more highly developed the more its elementary parts differ from one another.[16] Owen argued that "vegetative repetition," the multiplication of similar parts for the performance of the same actions, is the simpler and inferior condition compared to the assemblage of less numerous parts, each fitted for the performance of a particular function.[17] Milne Edwards said that perfection within each type is achieved principally by division of physiological labor, the restriction of particular functions to particular organs.[18] When in 1851 Lyell at great length reiterated his nonprogressionist interpretation of the fossil record, Owen, in an equally long rebuttal, explained these new views of what constitutes an "advance of organization" and urged that when the succession of organisms is judged by the criterion of degree of specialization of a general type form, it appears that there has been progress.[19]

In the 1850s increasing differentiation, specialization, or division of labor became the generally accepted definition of organic progress among zoologists. As such it was adopted as the basis for numerous interpretations of the history of life. Owen and Carpenter separately proposed the law of development from more general to more special forms as the best expression of the character of the geological succession of organisms.[20] Their new version of progressionism was accepted even by some of the older progressionist geologists, such as Sedgwick and Murchison.[21] Robert Chambers in the tenth edition of *Vestiges of the Natural History of Creation* (1853) revised his account of organic development to bring it into agreement with Owen's and Carpenter's theory.[22] And Herbert Spencer, inspired by Owen's lectures and Carpenter's *Principles of Comparative Physiology*, made von Baer's law of development the fundamental principle of his universal evolutionary synthesis: Von Baer's "law of organic progress is the law of all progress," he said, and he cited Owen and Carpenter in support of his claim that the history of life, like the history of the solar system, is characterized by increasing heterogeneity.[23]

In Darwin's notes and correspondence from the mid-1840s on, his growing acquaintance with and approval of the new views on how to define progress are apparent. By 1845 he was aware of Owen's principle of vegetative repetition, but as the following passage shows, he was still not entirely free from the conviction that at least among the vertebrates man is the goal of progressive development. His belief that all animals are perfectly adapted seems also to have made it difficult for him to think in terms of degrees of perfection:

> What is the *highest* form of any class? Not that which has undergone most changes, for changes may reduce organisation. – generally however, that which has undergone most changes & which approaches nearest to Man? – Hardly applicable to insects or plants – Each perfect for its end, so not most perfect – "complication not of homologous organs" "combined, when comparison [possible], with Man as a model" Perfect adaptation comes in idea.[24]

For some time perfect adaptation remained a consideration for Darwin, but like most of his contemporaries he increasingly tended to separate the question of adaptation from that of degrees of perfection in comparison with the type of the group.[25] In the spring of 1846 Darwin discussed vegetative repetition with Owen and then with Hooker, indicating to the latter his favorable opinion of the idea.[26] Later that year he read two articles by Milne Edwards in which division of physiological labor was put forth as the means by which organization advances. In his notes on one of these, "Observations sur la Circulation" (1845), he attempted to explain Owen's law that highness is characterized by a decrease in the number of homologous parts.[27] From Milne Edwards's 1844 essay on classification he quoted the passage on division of labor and then appended a number of reflections on highness and lowness and on his theoretical expectations:

> (The idea of intellect & brain in man breeds much confusion: remember that Owen says that osseous fishes are most ichthistic, but the cartilaginous fishes are higher, if we look to affinity to higher tribes: the cartilaginous are oldest) p. 77 [Milne Edwards] remarks on the difficulty caused by development in some cases making organisms less complicated, as in Lernea (which I sh^d think was the strongest case known. Barnacles in some sense, eyes & locomotion, are lower, but then so much more complicated, that they may be considered as higher) . . . we then see that highness does not depend on perfection & number of organs, but on development . . . If we consider the number of changes as *highness*, then Lernea, a mere reproductive sack, w^d be higher; this is too revolting to common sense . . . It is evidently most difficult to make out which is higher & so no wonder little accordance with Geolog. History.

Darwin expected an "accordance" between degree of development and time of appearance (geological history). But his conclusion here, as in his notes on Chambers's *Vestiges*, was that the intricacy of the question made it advisable to avoid the expressions "higher" and "lower" and instead to state his argument in terms of embryonic development. Development is sometimes retrograde, as in the case of barnacles, but usually the adult is more complicated and perfect than its embryo: "all that we ought to expect is that if our fossils were perfect, that embryonic forms were the oldest; & hence in this discussion, leave out the terms higher & lower."[28]

Darwin returned to these themes in the summer of 1845 when he and Hooker corresponded on the definitions of "highness" and "lowness." Because of the difficulties caused by retrograde development and by the weakness of the fossil evidence, Darwin said still that all he expected was that ancient forms should "tend to resemble the larval or embryological stages of the existing." But he supposed that development generally results in an increase in perfection. Having read Huxley's translation of von Baer (1853), he told Hooker, "I am inclined to think that 'highest' usually means that form which has undergone most 'morphological differentiation' from the common embryo or archetype of the class." He thought the best definition, however, was "the specialisation of parts to different functions, or 'the division of physiological labour' of Milne Edwards."[29] Despite some lingering reservations, Darwin by this point was fairly well satisfied that Owen, von Baer, and especially Milne Edwards had formulated useful definitions of progress, and he was willing to make them the basis for his own speculations.[30]

Between 1854 and 1857 the principle of divergence and the associated transformation of Darwin's theory gave him two explanations of progress whose combined effect was to resolve the tension between adaptation and development. Their import was that the improvement of adaptation is an ongoing rather than an inherently intermittent process and that improvement of adaptation is usually effected by an increase in specialization and complexity. One of these explanations is that according to the theory of natural selection, organisms will always be "higher" than their predecessors in the sense that they must have had some advantage over their predecessors in order to beat them in the struggle for life. Because the adaptedness of a form is relative to that of other organisms with which it competes, the degree of "highness" of a species or group of species will depend on the severity of the competition that has produced it. "Species inhabiting a very large area, and therefore existing in large numbers, and which have been subjected to the severest competition with many other forms, will have arrived, through natural selection, at a higher stage of perfection than the inhabitants of a small area." Since Europe and Asia are the largest territory, their productions' are the "most 'improved,'" while those of Australia are "less-perfected." Ever-present competition means that there is a tendency to change and to this kind of improvement. But where the competition is less severe there will be less improvement, which accounts for anomalies – living fossils, such as *Ornithorhynchus* (an inhabitant of Australia) and *Lepidosiren* (an inhabitant of fresh water).[31]

Darwin called this kind of improvement "competitive highness," and he insisted that it necessarily followed from his theory: "On our theory of Natural Selection, if the organisms of any area belonging to the Eocene or Secondary periods were put into competition with those now existing in the same area (or probably in any part of the world) they (*i.e.* the old ones) would be beaten hollow and be exterminated; if the theory be true, this must be so." In 1858, when this was written, a tendency toward "competitive highness" was indeed a necessary consequence of natural selection. But it became such only as a result of the changes in Darwin's theory that occurred during the previous four years. It is not mentioned in the "Essay of 1844" because, as Darwin then conceived of the evolutionary process, it entailed no such tendency. Darwin in 1844 thought species were perfectly, rather than relatively well, adapted, and he thought of adaptation as having reference primarily to physical conditions. On such a view, Eocene forms were as perfectly adapted to their world as present forms are to theirs. If a competition between them were held under present conditions, the Eocene forms would be beaten, but if under Eocene conditions, the modern forms would be exterminated.[32]

The idea of "competitive highness" depends less on the theory of natural selection than on the concept of relative adaptation, and consequently it became part of Darwin's theory only in the mid-1850s, when he began to think of evolution as potentially continuous and virtually independent of external change. Although it is not then a necessary adjunct of natural selection, it is a peculiarly Darwinian conception of progress, one that makes sense only in the context of Darwin's theory. For this reason it was not a fully satisfactory solution to Darwin's long-standing problem of explaining progress in the more usual senses of the word. But this also he was able to do as a result of his new understanding of the evolutionary process. The same considerations that led Darwin to his principle of divergence convinced him that natural selection will tend to produce not only competitive highness, but also increasing specialization or division of labor. This is because specialization appeared to be an advantage in the struggle for existence. Implicit in all of the new definitions of highness, but most obviously so in that of Milne Edwards, was the claim that specialization is beneficial.[33] It was this feature of the new definitions that made it possible for Darwin to explain at last why natural selection will result in progress. Since specialization is an advantage, organisms that are specialized for diverse modes of existence will tend to be selected.

In discussing divergence Darwin made no distinction between diversification of a group of related species and the division of physiological labor in the individuals of each species of the group. He treated them as two aspects of the same tendency toward specialization. This is most apparent in the examples with which he illustrated the claim that his principle of divergence was the same doctrine as Milne Edwards's division of labor:

> Let us take an imaginary case of the Ornithorhynchus; & suppose this strange animal to have an advantage over some of the other inhabitants of Australia, so as to increase in numbers & to vary: it could, we may feel pretty sure, increase to any *very great* extent, only by its descendants becoming modified, so that some could live on dry land, some could feed exclusively on vegetable matter in various stations, & some could prey on various animals, insects, fish or quadrupeds. In fact its descendants would have to become diversified, somewhat like the other Australian marsupials, which, as Mr. Waterhouse has remarked, typify in their several sub-families, our true carnivores, insectivores, ruminants & rodents. Moreover it can, I think, hardly be doubted, that these very marsupials would profit by a still further division of physiological labour; that is by their structure becoming as perfectly carnivorous, ruminant & rodent as are our old-world forms; for it may well be doubted (not here considering the probable intellectual infirmity of the marsupialia in comparison with the other or placentate mammals) whether many marsupial vegetable feeders could long exist in free competition with true ruminants, & perhaps still less the carnivorous marsupials with

true feline animals. And who can pretend to say that the mammals of the old world are diversified & have their organs adapted to different physiological labours to the extreme, which would be best for them under the conditions to which they are exposed? Had we known the existing mammals of S. America alone, we should no doubt have thought them perfect & diversified in structu*r*e & habit to the exact right degree; but the vast herds of feral cattle, horses, pigs & dogs, at least show that other animals, & some of them as the horse & solid-horned ruminants, very different from the endemic S. American mammals, could beat & take the place of the native occupants.[34]

Here we see displayed the intimate connection in Darwin's mature theory between divergence, "competitive highness," and progress. The tendency he saw toward increasing diversification was at the same time a tendency toward competitive highness and increasing specialization of the majority of organisms. This is the basis for his statement to Hooker that "a long course of 'competitive highness' will ultimately make the organization higher in every sense of the word." In notes from the same period Darwin explicitly equated "best adapted" and "most divergent."[35] Division of labor is an advantage; therefore "natural selection will always tend, where habits permit, to specialise organs."[36]

Darwin had fully worked out his ideas on progress by 1857 and contemplated discussing them, including their bearing on human evolution, in *Natural Selection*. In a note ticketed for the treatment of "Difficulties" he wrote, "Embryology shows there has been advance Nat. Sele. tends to specialise & advance – so in intellect. I must enlarge on this."[37] But several constraints induced him to understate his argument in the first edition of the *Origin*.[38] The reception of *Vestiges* had long since suggested the advisability of avoiding any detailed discussion of the subject, and Huxley's recent review of Chambers's tenth edition indicated that the same kind of ridicule might be in store for him from this new champion of Lyell's nonprogressionism.[39] Darwin wanted Lyell's support and, at the least, Huxley's neutrality. Hooker too was sympathetic to Lyell's doctrines. On the other hand, any progressionists who might be favorably disposed toward Darwin's theory could be counted on to assume that transmutation would lead to progress. Darwin had nothing to gain and perhaps much to lose by exhibiting prominently his belief that progress is inevitable. Moreover, what might be called Darwin's scientific caution held him back. On theoretical grounds he was sure that progress must occur. But the arguments of Lyell, Hooker, Huxley, and others led him to think that the evidence for it was not as strong as he wished.

By the time of the second and third editions (1860, 1861) these constraints had lost much of their strength. From the first the *Origin*, in contrast to the *Vestiges*, was treated as a serious contribution to science. Darwin knew too that in "morphological differentiation" and "division of physiological labour" he had definitions of "highness" that were generally accepted among professional zoologists. Huxley remained committed to nonprogressionism for several more years; but though Darwin was skeptical about Huxley's competence to expound the theory of natural selection, he knew that he had his good will and support, so that he feared no public attack from that quarter. By contrast, Lyell, the founder and chief advocate of nonprogressionism, in 1859 abandoned his uniformitarian theory of the organic world.[40] Lyell's conversion to progressionism eliminated one of Darwin's reasons for avoiding the subject and, as it happened, supplied a perhaps decisive motive for treating it at greater length in subsequent editions. Lyell indicated to Darwin in October 1859 that he thought a continued intervention of creative power or some "principle of improvement" was necessary to produce successively higher levels of organization and, in particular, to produce man. Darwin responded quickly that natural selection alone is a sufficient explanation of improvement. Better adapted forms, forms better able to compete, more specialized forms, an increased number of forms, and more complex organic conditions are all inevitable consequences of natural selection, he said. "I can see no limit to this process of improvement, without the intervention of any other and direct principle of improvement." He added, "If I have a second edition, I will reiterate 'Natural Selection,' and, as a general consequence, 'Natural Improvement.' "[41]

Darwin made few changes in the second edition, but among them were two brief responses to Lyell. In the summary to the chapter on natural selection he inserted the remark that "it leads to the improvement of each creature in relation to its organic and inorganic conditions of life."[42] In his discussion of geological succession, where originally he had spoken only of competitive highness, he added that the best definition of highness is greater division of physiological labor and that, since this is an advantage to each being, natural selection "will constantly tend" to make later forms "higher" than their progenitors.[43] For the third edition Darwin drafted a whole new section, *"On the degree to which Organisation tends to advance,"* to follow his discussion of the principle of divergence. There he asserted, and repeated in all later editions, that "if we look at the differentiation and specialisation of the several organs of each being when adult (and this will include the advancement of the brain for intellectual purposes) as the best standard of highness of organisation, natural selection clearly tends towards highness."[44]

Darwin never abandoned his cautious approach to the question of progress. Whenever he said that natural selection tends to produce an advance in organization he made a point of insisting that there is "no necessary and universal law of advancement or development" and that it is likely that some simple forms will remain unchanged and that "retrogression of organisation" will occasionally occur.[45] But none of his readers or correspondents could have been in any doubt as to what he believed – that organization on the whole has advanced and is advancing still as a result of the selection of ever more highly specialized and diverse forms. "There is nothing in my theory," he wrote to the botanist W. H. Harvey, "necessitating in each case progression of organisation, though Natural Selection tends in this line, and has generally thus acted."[46]

Accident and inevitability

Darwin's efforts to develop a theory of progress can be explained in part by the functions his belief in improvement served in his evolutionary system. One was to justify the apparent evils of the Darwinian view of nature. In the conclusions to the "Sketch," "Essay," and *Origin* Darwin argued that famine, death, and the war of nature lead to "the most exalted object which we are capable of conceiving, namely, the production of the higher animals."[47] I noted earlier that Darwin's reasoning here is the same as that of Malthus and Paley on the evils of superfecundity, and probably the theological underpinnings of his argument and theirs were essentially the same: belief that the laws of nature were designed by a benevolent God. Maurice Mandelbaum has suggested that Darwin's belief in progress was simply a consequence of his theological convictions in the late 1830s, but I suspect that the situation was rather more complex, that there were nontheological underpinnings to the argument as well.[48] For one in Darwin's position there were plenty of reasons, other than theological ones, for believing in progress. And progress, apart from any explicitly theological considerations, was sufficiently widely perceived as good that it could serve, regardless of one's religious convictions, to justify struggle, inconvenience, misery, and the economic practices and rationalizations that helped produce them. The conclusion to the *Origin* might well have been not only a justification of the ways of God, the author of the laws of nature, but also a justification of the ways of the author of the theory, that is Darwin, who found a new, a painful and bloody nature, but a nature that was nevertheless good because it was progressive.[49]

Darwin's belief in progress served also to support his views on human evolution, past and future. John Greene has shown that Darwin thought natural selection has been an important, probably the most important, engine of human progress and that natural selection has recently acted, and continues to act, to eliminate the "lower" and preserve the "higher" civilized human races.[50] In his discussions of the descent and ascent of man, Darwin's general theory of organic progress was useful on several counts. The inevitability of progress meant that, given one or a few primordial created forms, the "higher" animals were bound to arise sooner or later.[51] And the tendency toward specialization implied to Darwin that if any animals appeared to whom something like a brain or mental faculties were at all important, natural selection could be expected ultimately to produce an intellectual being. In his explanations of the tendency toward specialization, Darwin was particularly concerned to point out its relevance to the question of the development of the brain and intellect. In a note for *Natural Selection,* already quoted, Darwin said natural selection "tends to specialise & advance – so in intellect." And when he introduced his long discussion of the advance of organization in the third edition of the *Origin,* he said that natural selection clearly leads to "the advancement of the brain for intellectual purposes."[52] In the *Descent of Man* he argued that if the moral and intellectual faculties were of high importance to man and his apelike progenitors, "they would have been perfected or advanced through natural selection." In a separate discussion on the origin of the moral sense, Darwin applied this same kind of argument – that once evolution has proceeded to a certain point, certain additional developments are inevitable. "The following proposition seems to me in a high degree probable," he said: "that any animal whatever, endowed with well-marked social instincts, the parental and filial affections being here included, would inevitably acquire a moral sense or conscience, as soon as its intellectual powers had become as well, or nearly as well developed, as in man."[53]

Here, as in other passages quoted above, Darwin uses the word "inevitably," and I have argued throughout, on the basis of Darwin's statements, that progress was for him a necessary outcome of the evolutionary process. But it may well be asked how natural selection, which involves so large an element of chance, can be said to lead inevitably to anything. This difficulty is perhaps partially responsible for the considerable disagreement among Darwin scholars (and others) about his theory of progress, or lack of one.[54] Darwin's doctrine of the differential probability of survival of chance variations does indeed rule out the possibility of predicting any particular event. One cannot say that man, as we know him, was an inevitable result of natural selection, and the same is true for every species and assemblage of species. But natural selection does not exclude the inevitability of various general results. If it did, the theory would be worthless. And one such inevitable result is the production of an organic world inhabited at all times by beings adapted well enough to survive. The inevitability of various general results is what is implied in Darwin's remark to Asa Gray that he looked at everything as the result of "designed laws" but with the details left to the working out of chance. The details are unpredictable. But the designer of the laws, as well as their discoverer, can predict the general ends which the laws will accomplish.[55] Darwin's statement to Harvey, quoted just above, is of the same sort. Natural selection does not necessitate "in each case" an advance of organization, but the overall result of natural selection is progress.

In Darwin's belief in the inevitability of certain general results of natural selection, such as adaptation and progress, there is an important point of contact between his thought and that of his contemporaries. For Owen, Carpenter, or Chambers, the development of life occurred according to a plan of creation. Every result, including man, was inevitable in the sense that it was preordained. For Darwin there was, after 1838, no such plan, but he did not reject the idea of inevitable results or the idea that these were the product of design. It is well known that although Darwin in his *Autobiography* (written in 1876) said he had gradually become an agnostic, he said also that when he wrote the *Origin* he was a "Theist." Other sources support his recollection, such as his letter to Hooker in 1870 in which he said he believed the universe was not the result of blind chance, though there is no evidence of design "in the details."[56] In this respect there is a considerable degree of continuity in Darwin's view of nature from the period of his earliest speculations until his old age. In his pre-Malthus notebooks, Darwin treated adaptation as part of a plan of creation; later he said it was an inevitable consequence of "designed laws." He originally thought man was designed, and progress too: "Has the Creator since the Cambrian formation gone on creating animals with same general structure. – Miserable limited view."[57] And in his autobiographical account of the religious views he held in 1859, he referred specifically to the existence of man as one reason for his belief in design. It is extremely difficult, or rather impossible, he said, to conceive "this immense and wonderful universe, including man with his capacity of looking far backwards and far into futurity, as the result of blind chance or necessity."[58] To make this consistent with natural selection it is necessary to assume that by "man" Darwin meant a moral and highly intellectual being, not necessarily man as we know him. Darwin's theory of progress made belief in the inevitability of such a being plausible. On the basis of his theory Darwin could suppose that, granted the evolution of creatures to whom the slightest amount of nervous activity was important, natural selection would inevitably improve some of their descendants until it had produced a being as intellectually advanced as man. And if, as is possible, the achievement of this level of specialization required first the development of the filial and parental affections, that is, social instincts, then these intellectual beings would inevitably be moral beings as well. In this sense Darwin's theory of progress went far, though perhaps not all the way, toward making a moral and intellectual being an inevitable – and therefore arguably designed – rather than a chance result of evolution.

Given Darwin's belief in the inevitability of progress, I can see no good reason for distinguishing sharply between his views and those of numerous other mid-Victorian speculators. Each theorist of progress had his own mechanism to propose, but the results they all imagined were remarkably similar – inevitable advance from the more general to the more special form. No single criterion will serve to separate Darwin's views from those of Owen, Chambers, Spencer, and others, short of saying simply that only Darwin believed progress was caused chiefly by natural selection. All attempts that I am acquainted with to make such a distinction are clearly inadequate. For instance, Jonathan Hodge, as well as others, has insisted that the only general criterion of progress Darwin recognized was competitive fitness.[59] As we have just seen, this is certainly wrong. Darwin adopted the criteria that were most widely accepted by his professional colleagues, and he asserted again and again that natural selection tends to produce what they called "highness." Michael Ghiselin has urged that for Darwin progress was "contingent."[60] But John Greene has pointed out that this was equally true for Spencer, and it was true in one sense even for Lamarck, despite his belief in an innate progressive tendency.[61] In the mid-nineteenth century all naturalists, most of whom believed in progress, admitted that retrograde development occurs. In embryology the barnacles were an obvious instance in which the adult seemed to be "lower" than the embryo, and in paleontology the reptiles as a group had clearly declined in the scale of perfection.

Stephen Gould has made much of Darwin's vow never to use the terms "higher" and "lower."[62] But Darwin consistently refused to adhere to his rule. He first adopted it in 1844 and 1845 when he read *Vestiges* and Sedgwick's review of it. But in the spring of 1846 he was discussing with Owen and Hooker what high and low mean. He repeated the vow in December 1846 when he read Milne Edwards, but over the next several years he frequently attempted to find a satisfactory definition of highness, he sketched numerous explanations of how highness is produced, he experimented with ways of introducing his intended discussion of highness in his species book,[63] he discussed definitions in his monograph on cirripedes,[64] and in his notes he freely used the terms "higher" and "lower." Darwin's exchange with Hooker in December 1858 I find particularly amusing and instructive. In a letter of December 24, Darwin explained his theory that the more severe the competition in a region, the "higher" the "stage of perfection" that the inhabitants will reach. When Hooker questioned his conclusions on the lowness of Australian plants, Darwin said, "I do not think I said that I thought the productions of Asia were *higher* than those of Australia. I intend carefully to avoid this expression." He then went on, however, to explain his view of "competitive highness," and by the end of the letter he was saying that a long course of competitive highness will make the organization "higher in every sense of the word."[65] Darwin imposed his rule of avoiding the term "higher" because of the difficulties involved in defining it, not because he had any objection in principle to the concept of organic progress. And he repeatedly broke his own rule because he wanted very badly to be able to say that natural selection tends to produce an advance in the organization of the majority of beings throughout the world. By 1860 he was saying just that and using, once again, the expression "higher."

If Darwin's nineteenth-century readers generally assumed that he, like Spencer (or Chambers), was a progressionist, it is easy to understand why.[66] He was. He saw the paleontological work of progressionists like Owen as supportive of his theory and wished that the evidence were stronger still. By contrast, he had no use for the paleontology of Lyell and Huxley. When Huxley in 1862 argued that Darwin's was the only transmutation theory compatible with nonprogression, Darwin did not thank him for associating natural selection with nonprogressionism or for distinguishing cleverly between Darwin's views and Lamarck's.[67] Darwin may have been more cautious than many in his pronouncements on progress. But with the introduction into his theory of the principle of divergence – of increasing division of labor – he, like Spencer, portrayed a dynamic, expanding, and advancing organic world. Without ceasing to be a theory of adaptation, natural selection became also a theory of development. The same developmental conceptions that informed the biological thought of most of Darwin's contemporaries found expression in the *Origin of Species*, not only in the chapters of Part II, but in Chapter IV, the heart of his discussion of natural selection.

Conclusion: The development of Darwin's theory as a social process

The theoretical work of a single scientist, in contrast to the formation and characteristics of an institution, a discipline, or a research school, is not an obviously attractive subject for a study whose goal is to reveal one aspect or another of the social character of science. Nor was this study initially undertaken for such a purpose. From the outset I have been less interested in the nature of science in the abstract than in the development of one particular theory. But as I proceeded I became more and more convinced that that development can best be described as a social process, and it is this idea that I want to discuss briefly in conclusion.

Since the history of science must inevitably concern itself at many points with scientific ideas, there will continue to be a need for the study of individual scientists who produce them. Such studies are likely to be more interesting and more useful as historians learn better how to recognize some of the ways in which the thought of individuals, even those who, like Darwin, have few if any formal professional and institutional ties, is shaped by the social and cultural conditions in which they work. The task will be easier in some cases than in others; and although that of Darwin seems to me to fall at the less difficult end of the scale, I do not pretend to have gone very far toward its completion here. One modest conclusion, however, appears to be fully justified: that the formation and transformation of Darwin's theory represent not so much the results of an interaction between the creative scientist and nature as between the scientist and socially constructed conceptions of nature. Even when Darwin was in closest contact with nature, as on the voyage and during his research on barnacles, his interaction with nature was mediated by assumptions and ways of perceiving nature that he derived from other naturalists, both his predecessors and his contemporaries, and from the culture in which he was educated and carried out his work. Three respects in which this is so may be specified. The first concerns Darwin's early belief in the harmony of nature and his long-held assumption that adaptation is perfect; the second, the conversion of natural selection into a theory not only of adaptation but of progressive development; and the third, Darwin's efforts in Part II of his species work to recast the generalizations of his contemporaries into evolutionary form.

Most of the efforts to link Darwin's science to the dominant ideological interests of early and mid-nineteenth century Britain have focused on the relationship of the theory of natural selection to the political-economic thought of his day, particularly to Malthus's theory of population. Malthus's theory, which for him and others served readily identifiable social and political interests, unquestionably suggested to Darwin the idea of natural selection. Do we then say that the interests that produced Malthus's theory also produced Darwin's? Perhaps. But there are other, more subtle ways in which scientific ideas are socially constructed, ways that involve not the direct transfer into nature of explicitly political theories, but rather the establishment in a culture, and the persistence over relatively long periods of time, of ways of seeing nature that are themselves constructed in response to social, political, and religious, as well as scientific, interests. At various stages in Darwin's career his belief that adaptation is perfect, that nature is a harmonious system of purposeful laws established by God, and that even change serves to maintain order and harmony, played key roles in his thinking and shaped in definite ways his theories of transmutation. These beliefs provided a useful framework for the investigations carried out by biologists of the first half of the nineteenth century, including Darwin's. As has frequently been noted, they posed the problem of adaptation for which Darwin offered natural selection as a solution.[1] But for natural theologians and political economists their main functions were religious, social, and political – to demonstrate the existence, wisdom, and goodness of God and to show that the existing social and political order is God's order.[2] More pertinent to the point I want to make is that the view of nature of which these beliefs were a central part not only was sustained in Darwin's day by the various interests it served, but was originally constructed in the late seventeenth and early eighteenth centuries as part of the ideology of dominant segments of the social, political, and religious elite in England.[3] Long before Darwin opened his first transmutation notebook, the elaboration by naturalists and others of ideas made prominent by natural theologians since the time of Boyle and Ray had made perfect adaptation and the purposiveness of variation "obvious characteristics of the natural world." Darwin accepted them as a matter of course, and they gave the early theory of natural selection the structure of a mechanism to preserve harmony.

This process can be described in more general terms. If people with particular interests succeed in making their view of the world appear "natural," theorists in succeeding generations, who may or may not have precisely the same interests, but who are located and educated in a society that has learned to see nature in the same way, will be very likely to produce theories that are shaped by assumptions and perceptions that originated in the political and social interests of their predecessors. And such theories, whatever the attitudes or intentions of their authors, will inevitably convey some parts of the ideology that is built into society's view of nature.[4] The advantage of looking at the question in this way is that it avoids the difficult problems of discovering the social, political, and religious interests of the scientist or of the social group to which he belongs and of correlating them directly with his scientific ideas.[5]

A similar process is apparent in Darwin's elaboration in the 1850s of a theory of progressive development. There is good textual evidence to suggest that Darwin's belief in progress and his desire to find an explanation for it were related to his religious views, his ideas on human evolution and the history of civilization, and his hopes for the future. But even if this evidence did not exist, there is much that could be said about the role of nonscientific interests in the construction of his theory of progress. Without attempting to answer the question of the reasons for Darwin's belief in progress or of the extent to which he was moved by the same social and political concerns that made faith in progress attractive to many other British intellectuals of his generation, we can with some confidence say that the idea of development that characterizes much of mid-nineteenth-century thought, both scientific and other, became a part of the theory of natural selection around 1857 and was publicly taught by Darwin in the *Origin of Species* from the second edition on. I am not competent to specify all the sources of this developmental outlook. In Great Britain it probably owed something to the introduction of German historical and biological thought, something to optimistic perceptions of recent material and political changes, and something to new ideological needs generated by these changes.[6] Whatever its sources, however, it is clear that it was not simply an immediate consequence of detached observation of nature: It was not until the second quarter of the nineteenth century that sizable numbers of British scientists who observed nature found in it developmental patterns.

In the 1830s Darwin, Owen, and Carpenter agreed on the inadequacy of strictly teleological explanations of organisms, and this opened the possibility of conceiving the history of life as a development from one or several simple type forms, which Darwin did almost immediately, Owen and Carpenter during the next decade. Owen and Carpenter took as their problem the discovery of laws of development and assumed that when they were discovered they would be found to produce the adaptation we see in nature. They were confident of this because they believed that these laws were part of God's unfolding plan of creation. Consequently they directed their speculations toward theories of organic development, with little regard for explaining adaptation. Darwin on the other hand began very early to look for a theory of adaptation. As a result, despite his belief that the history of life has been predominantly upward divergence from one or a few simple ancestors, Darwin's theory was not at first a theory of development. Rather than a law of development from the general to the specialized form, such as Owen and Carpenter produced by ignoring the problem of adaptation, Darwin described an intermittent process of adjustment to environmental change. But Darwin made use of developmental concepts supplied by his contemporaries, and in his efforts to explain aspects of their developmental approach to natural history, he incorporated some of these concepts – most notably Milne Edwards's definition of degree of development – directly into the theory of natural selection. With the addition of the principle of divergence, natural selection became a theory of development as well as of adaptation. In the second and subsequent editions of the *Origin*, Darwin argued that the equation "more fit" = "higher" is generally correct. Those who later took Darwin's theory as a basis for ideologies of progress and of the natural dominance of "higher" over "lower" human races were not required to distort it to suit their purposes. The theory itself already contained elements of such an ideology. Again, this says nothing about Darwin's own attitudes or intentions. It says only that he was building his theory out of ideologically loaded concepts.

We should, I think, look at this change in Darwin's theory as a microcosm of the more general development from a philosophy of nature and man appropriate for an agrarian and aristocratic world to one suitable for the age of industrial capitalism.[7] But the shift in Darwin's theory does not appear to reflect any alteration in his own social and political outlook. He believed in progress in 1837 and in 1857, apparently for the same reasons. The fact that in 1857 he found himself in possession of a theory of progress was largely due to what he probably saw simply as the scientific demands of his speculative work – the need to explain aspects of contemporary natural history – and to his use of current scientific concepts. And the relationship of these concepts to the general shift toward developmental ideas about nature and society is undoubtedly more apparent now than it was to Owen or Milne Edwards or Darwin. It may be that it is usually through such labyrinthine and obscure pathways as Darwin's speculations followed en route to his principle of divergence and theory of progress – pathways seemingly so insulated from society and so "internal" to Darwin's scientific interests – that economic and social relations and their changes shape scientific thought.

At another level the development of Darwin's theory suggests how the work of an individual scientist in the privacy of his study or laboratory – even a scientist who is engaged in creating a revolutionary theory designed to supplant existing explanations – is molded by the ideas and attitudes of his professional colleagues. Darwin's contemporaries, believing in a plan of creation or in the purposive development of the world spirit, found organic nature to be exceedingly regular, and they proposed sweeping generalizations expressive of this regularity, such as von Baer's law of embryonic development and Milne Edwards's exact correlation between the divergence of embryos and the affinities of adults. Accepting their generalizations as facts, Darwin interpreted them as indicating tendencies in the evolutionary process. Mere chance could not produce such order. By developing explanations for them, he incorporated into his theory elements that embodied his contemporaries' predilection for seeing orderly development in nature. One such element is the principle of divergence. From the perspective of many twentieth-century evolutionary biologists, this may look like a retrograde step for Darwin;[8] but from the perspective of the development of Darwin's thought up to the mid-1850s, it appears as the catalyst that produced the concept of relative adaptation, which marks the culmination of Darwin's gradual rejection of most of the natural theological assumptions about nature that he initially shared with his contemporaries. His departure from their view of nature was conditioned by his efforts to explain their science.

The generalizations of Darwin's contemporaries did not of course determine in any rigorous fashion the shape his theory would have. Scientific speculation is an unpredictable business. Given the biological literature of Darwin's day, his theory might easily have developed along paths other than those actually taken. There was no necessity for him to work out a principle of divergence, for instance. On the other hand, had the ideas of his fellow naturalists been different, his theory would certainly have developed in different ways. If his colleagues had proposed no regular laws of development for Darwin to explain, natural selection might well have retained the structure it had in the "Essay of 1844."

The rejection of teleological explanation, the rejection of static conceptions of nature in favor of developmental conceptions, and the dramatic increase in knowledge about embryonic development, organic structure, the fossil record, and geographical distribution that occurred in the mid-nineteenth century can be called without exaggeration a quiet revolution taking place around Darwin while he worked on his theory of natural selection. Its authors were his professional colleagues, with whom he long shared many basic assumptions. Its new ideas made a substantial contribution to Darwin's evolutionary synthesis. And the problems it posed for Darwin stimulated some of his most important theoretical work. Where this revolution might have led had the *Origin of Species* not intervened to produce the Darwinian revolution, it is not possible to say – probably to various theories of descent without natural selection, such as were in fact widely held by biologists within a decade of 1859. In any event, it was the work Darwin was doing in London and at Down, the construction of a new theory and the detailed demonstration of how contemporary natural history could be given a new and interesting form, that made evolution scientifically respectable and secured its widespread acceptance. But that work was intimately bound up with the main currents of mid-nineteenth-century biological thought.

Notes

The following abbreviated titles are used throughout the notes:

B notebook, C notebook, D notebook, E notebook
> Gavin de Beer, M. J. Rowlands, and B. M. Skramovsky, eds., "Darwin's Notebooks on Transmutation of Species," *Bulletin of the British Museum (Natural History), Historical Series,* 2 (1960): 23–183; "Pages Excised by Darwin," ibid., 3 (1967): 129–76. Darwin's B, C, D, and E notebooks are De Beer's First, Second, Third, and Fourth. Reference throughout is to Darwin's page numbers; "e" designates an excised page: e.g., B notebook, p. 29e.

"Essay of 1844," "Sketch of 1842"
> Francis Darwin, ed., *The Foundations of the Origin of Species: Two Essays Written in 1842 and 1844* (Cambridge: Cambridge University Press, 1909).

LLD
> Francis Darwin, ed., *The Life and Letters of Charles Darwin,* 2 vols. (New York: D. Appleton, 1888).

MLD
> Francis Darwin and A. C. Seward, eds., *More Letters of Charles Darwin,* 2 vols. (New York, D. Appleton, 1903).

Natural Selection
> Robert C. Stauffer, ed., *Charles Darwin's Natural Selection* (Cambridge: Cambridge University Press, 1975).

Origin
> Charles Darwin, *On the Origin of Species* (London: John Murray, 1859).

Reference numbers to manuscripts in the Darwin Archive at Cambridge University Library, e.g., DAR 205.9:79, denote Darwin volume (205.9) and folio, letter, or item number (79). Where the document is a manuscript with pages numbered by Darwin, these numbers are denoted as DAR 30.1:MS p.65. Comments on books in Darwin's Library refer to works in the Darwin Archive.

Introduction: Darwin and his fellow naturalists

1 There have been several studies of this important early period. See especially Sandra Herbert, "The Place of Man in the Development of Darwin's Theory of Transmutation," Parts 1 and 2; David Kohn, "Theories to Work By"; Camille Limoges, *La Sélection Naturelle.* See also Howard E. Gruber, *Darwin on Man:* Malcolm J. Kottler, "Charles Darwin's Biological Species Concept and Theory of Geographic Speciation."

2 On Darwin's life the best source is still *LLD*. See also Gruber, *Darwin on Man:* Gertrude Himmelfarb, *Darwin and the Darwinian Revolution,* pp. 1–146; [Geoffrey Harry Wells] Geoffrey West, *Charles Darwin.*

3 See for instance W. Faye Cannon, "The Whewell-Darwin Controversy," p. 383; Stephen J. Gould, *Ever Since Darwin,* pp. 21–7; Sylvan S. Schweber, "The Origin of the *Origin* Revisited," pp. 310–15.

4 In considering the first, Derek Freeman has given a brief answer to the second – that Darwin spent the years collecting needed evidence: "The Evolutionary Theories of Charles Darwin and Herbert Spencer," pp. 232–3.

5 Throughout I use the terms "evolution" and "evolutionary," despite the fact that Darwin did not employ them before 1859. There is some slight warrant for this in Darwin's use of the word "evolved" in his essays of the 1840s: "Sketch of 1842," p. 52; "Essay of 1844," p. 255. See Peter J. Bowler, "The Changing Meaning of 'Evolution.'"

6 See *Natural Selection,* "General Introduction."

7 Whether the harmony was self-adjusting or required the intervention of the creator to maintain it amid the changes revealed by geology was a question on which thinkers might differ who nevertheless agreed on the fact of harmony.

8 The two best accounts of the formation of Darwin's theory deny it. See Kohn, "Theories to Work By," and Limoges, *La Sélection Naturelle.* Ernst Mayr hints at it in "Darwin and Natural Selection," p. 326.

9 Jerome R. Ravetz's comment on the social character of the work of the individual scientist is worth recalling in this regard: "The personal endeavour that is necessary for worthwhile scientific work is itself a very artificial and social creation. For the individual talent and style, and the special private knowledge, are applied in a highly stylized fashion: to the investigation of problems concerning a given set of intellectually constructed objects, working up the materials derived from experience in accordance with established methods, and drawing conclusions within accepted patterns of argument. Even the greatest creative work, in which all these components may be strongly modified, must base itself on a tradition in which such modifications themselves are a natural development" (*Scientific Knowledge and its Social Problems,* p. 237).

10 In terms of competence, career orientation, and audience to which he

proposed to speak. On professionalism, with specific reference to Darwin, see Herbert, "Place of Man," Part 2, pp. 157–8; and Susan F. Cannon, *Science in Culture*, pp. 137–65.

11 Milne Edwards has fared slightly better than the others in this respect. See Limoges, *Sélection Naturelle*, pp. 135–6, and "Darwin, Milne-Edwards et le Principe de Divergence."

12 The relation of the political-economic thought of the early nineteenth century to Darwin's science has been explored in numerous works. See especially the studies on Darwin and Malthus cited in Chapter 3, n. 3.

1. Darwin and the biology of the 1830s

1 In the biological literature of the mid-nineteenth century, "teleologist" means "one who seeks to explain the phenomena of organic nature solely by reference to final causes." As will become clear in the following discussion, antiteleologists believed in a created, purposeful universe. Their rejection of teleology was similar to the rejection of teleological explanations in physics in the seventeenth century, when also the purposiveness of the whole was rarely disputed. For examples of mid-nineteenth-century usage see [William B. Carpenter], "Physiology an Inductive Science," p. 340; William B. Carpenter, *Principles of General and Comparative Physiology*, p. 561: "The Teleologist (who rests satisfied with the evident *object* of this adaptation as a sufficient *reason* for its occurrence) . . . "; Richard Owen, *On the Nature of Limbs*, pp. 10, 84.

2 *Origin*, p. 201.

3 Georges Cuvier, *Le Règne Animal*, 1:16–17.

4 Ibid., p. 6.

5 Toby Appel, "The Cuvier-Geoffroy Debate and the Structure of Nineteenth Century French Zoology," esp. pp. 109–25. Given the actual arguments and practice of Cuvier and other teleologists, it is difficult to accept Russell's view that, following Kant, they used teleology as merely a regulative principle. E. S. Russell, *Form and Function*, p. 35.

6 Etienne Geoffroy Saint-Hilaire, *Philosophie Anatomique*, 1:xxv–xxviii; and *Principes de Philosophie Zoologique*, pp. 3, 51.

7 Geoffroy, *Philosophie Zoologique*, pp. 63–6; Russell, *Form and Function*, pp. 74–8.

8 William Buckland, "Lecture," pp. 104–5; William Whewell, *History of the Inductive Sciences*, 3:444–78; Charles Bell, *The Hand*, pp. 41, 153–62, 280.

9 Before 1830 there were of course many biologists who were not committed to the strictly teleological approach (Geoffroy, Lamarck, most of the German morphologists), but in the 1830s there was a conscious movement against teleological explanation, especially in Britain, where it had been most insisted on. For the situation in France before (as well as after 1830, see Appel, "The Cuvier–Geoffroy Debate"; see also Pietro Corsi, "The Importance of French Transformist Ideas for the Second Volume of Lyell's *Principles of Geology*."

10 Appel, "The Cuvier–Geoffroy Debate," has given some attention to this

for the French biologists she considers. Carpenter's concern to claim the same status for physiology as physics enjoyed will be apparent in the discussion below.

11 A similar point is made in Peter Bowler, *Fossils and Progress*, pp. 36–9.

12 Some of these biological theories are discussed in Chapter 5, and Darwin's recasting of them in Chapters 6 and 7.

13 Owen, *On the Nature of Limbs*, pp. 9–10, 40.

14 Bell, *The Hand*, pp. 42, 153–61, 280.

15 Peter Mark Roget, *Animal and Vegetable Physiology considered with reference to Natural Theology*, 1:48–50; 2:626–8.

16 Martin Barry, "On the Unity of Structure in the Animal Kingdom," p. 116.

17 Ibid., pp. 362–4.

18 Richard Owen MSS, Hunterian Lectures for 1837, Royal College of Surgeons.

19 William Whewell, *Astronomy and General Physics Considered with Reference to Natural Theology*, p. 353.

20 Whewell, *History of the Inductive Sciences*, 3:464, 470.

21 See J. D. McFarland, *Kant's Concept of Teleology*, pp. 113–16, 123–6; see also H. W. Cassirer, *A Commentary on Kant's Critique of Judgment*, pp. 354, 357–60; Immanuel Kant, *Kritik der Urtheilskraft*, 5:397–401.

22 Whewell, *History of the Inductive Sciences*, 3:470 (Kant, *Kritik der Urtheilskraft*, p. 376 [section 66]).

23 McFarland, *Kant's Concept of Teleology*, pp. 97, 137. See also Cassirer, *Commentary on Kant's Critique of Judgment*, p. 356.

24 By 1853 the further progress of biology had helped persuade Whewell that the teleological approach was inadequate. See Chapter 5, pp. 142–3.

25 [Carpenter], "Physiology an Inductive Science," pp. 321, 328–29.

26 Ibid., p. 338 (quoting Whewell, *Astronomy and General Physics*, p. 353).

27 William B. Carpenter, *Principles of General and Comparative Physiology*, p. 461. A parallel passage occurs in "Physiology an Inductive Science," p. 340.

28 It is hardly necessary to point out that most of them held Lamarck's explanation of adaptation to be both inadequate and improbable.

29 See Dov Ospovat, "Lyell's Theory of Climate." On the concerns that induced Lyell to construct an antievolutionary system, see Michael Bartholomew, "Lyell and Evolution."

30 Leonard G. Wilson, ed., *Sir Charles Lyell's Scientific Journals on the Species Question*, p. 6. See also Charles Lyell, *Geological Evidences of the Antiquity of Man*, p. 422.

31 Charles Lyell, *Principles of Geology*, 1:146–7. Unless otherwise specified, all references to Lyell's *Principles* are to the first edition.

32 A. P. De Candolle, *Essai Elémentaire de Géographie Botanique*, p. 44. Lyell, *Principles*, 2:70–72.

33 De Candolle, *Essai*, p. 44; Lyell, *Principles*, 2:72.

34 De Candolle, *Essai*, pp. 33–4. Near the conclusion to his essay, De Candolle restated this same idea: "Species are distributed over the globe in part according to laws which one may immediately deduce from the

combination of the known laws of physiology and physics, [and] in part according to laws which appear to relate to the origin of things and which are unknown to us" (p. 61).

35 Wilson, ed., *Lyell's Scientific Journals*, p. 5. Lyell, *Principles*, 1:123. This point is discussed more fully in Ospovat, "Lyell's Theory of Climate." Camille Limoges has also discussed, in somewhat different terms, the distinction between De Candolle's views of nature and views such as Lyell's: *La Sélection Naturelle*, esp. pp. 59–69.

36 See Martin J. S. Rudwick, "Uniformity and Progression"; R. Hooykaas, *Catastrophism in Geology;* Philip Lawrence, "Charles Lyell versus the Theory of Central Heat."

37 Lyell, *Principles*, 1:123.

38 Martin Rudwick has called this complex of ideas the "directionalist synthesis." See his "Uniformity and Progression," pp. 213–17.

39 Henry T. De la Beche, *Researches in Theoretical Geology*, p. 240; William Buckland, *Geology and Mineralogy Considered with Reference to Natural Theology*, 1:107; Adam Sedgwick, *Discourse on the Studies of the University of Cambridge*, pp. lv–lvi.

40 Carpenter, *Physiology* (1st ed., 1839), p. 177.

41 A still useful discussion of this and other ideas of a parallel between geology and biology is R. Hooykaas, "The Parallel between the History of the Earth and the History of the Animal World."

42 Louis Agassiz, *Essay on Classification*, pp. 91–94.

43 Ibid., pp. 15–23, 140–52, 199–206. Similar arguments against environmental determinism were used by James C. Prichard, *Researches into the Physical History of Mankind*, pp. 50–1.

44 Owen, *On the Nature of Limbs*, p. 86; [Richard Owen], "Generalizations of Comparative Anatomy" (by relationship he meant structural similarity).

45 Sedgwick, *Discourse*, pp. ccxiii–ccxiv. See John H. Brooke, "Natural Theology of the Geologists," pp. 45–7.

46 Agassiz, *Essay on Classification*, pp. 12, 17.

47 Owen, *On the Nature of Limbs*, p. 86.

48 Carpenter, *Physiology* (1st ed., 1839), pp. 463–4.

49 Carpenter, *Physiology* (2nd ed., 1841), p. 192: "If any number of living beings had come into existence, without that adaptation to their conditions of existence which we observe in those now living, they would long ago have disappeared from the surface of the globe. In fact, it has been from changes in the external conditions to which they had not the power of conforming, that many races *have* become extinct. It *might* be argued, then, that the cases in which we observe this adaptation are only those in which it chanced to exist, out of a much larger number in which it was deficient; and, however *improbable* such a supposition may be, it would not be easy to prove its *impossibility*." (Carpenter of course did not call this natural selection.)

50 Richard Owen MSS, Hunterian Lectures for 1838, Royal College of Surgeons.

51 Owen, *On the Nature of Limbs*, pp. 84–6.

52 Theodor Schwann, *Microscopical Researches into the Accordance in the Structure and Growth of Animals and Plants*, pp. 187–8.

53 Ibid., p. 187.

54 With respect to geographical distribution, see Darwin's comment in the *Origin*, p. 346: "Neither the similarity nor the dissimilarity of the inhabitants of various regions can be accounted for by their climatal and other physical conditions. Of late, almost every author who has studied the subject has come to this conclusion." On zoologists generally, see Mary P. Winsor, *Starfish, Jellyfish, and the Order of Life*, pp. 139–41, 177–8. For the pro-Geoffroy attitude in France after 1830, see Appel, "The Cuvier–Geoffroy Debate." On von Baer and Milne Edwards, see Chapter 5.

55 [Owen], "Generalizations of Comparative Anatomy," p. 80.

56 This first occurred to me as a result of reading David Kohn's study of the transmutation notebooks, where it is argued that Darwin did not give up, but rather modified, the idea of perfect adaptation in the period before he read Malthus: "Theories to Work By," pp. 98–9, 104–5, 143.

57 Charles Darwin, *The Autobiography of Charles Darwin*, pp. 59, 87.

58 Three of the most important of these have been noted by Sandra Herbert in "The Place of Man in the Development of Darwin's Theory of Transmutation," Part 1, pp. 233–6.

59 Nora Barlow, ed., *Charles Darwin's Diary of the Voyage of H.M.S. "Beagle,"* p. 383; Charles Darwin, *Journal of Researches into the Geology and Natural History of the Various Countries visited by H.M.S. Beagle, pp.* 526–7.

60 DAR 30.1:MS p. 65.

61 DAR 30.2:MS p. 200.

62 DAR 42 (ser. 3):MS p. 2, in geological notes dated February 1835. In an apparently earlier set of notes, Darwin expressed similar ideas on the close relationship between environmental conditions and the organic creation: "It would appear to be necessary under similar circumstances, that the landscape should possess the same forms & tints" (DAR 30.2:MS p. 156).

63 DAR 42 (ser. 3):MS p. 7, in "Reflections on reading my Geological Notes."

64 Ibid., MS p. 10. See also *LLD*, 1:363.

65 "Creation does not bear upon solely adaptation of animals" (Nora Barlow, "Darwin's Ornithological Notes," p. 277). On Darwin's notebook R.N., from which this passage is quoted, on the dating of the relevant transmutationist passages, and on Darwin's adoption of transmutationist views, see Herbert, "The Place of Man, Part I," pp. 233–49.

66 Darwin was pleased when Lyell and John Herschel alluded in letters to "intermediate causes" of the origination of species; but much of his argument against "creation" in the notebooks was directed toward Lyell's teleological account of the appearance of new forms. Of course Darwin was also opposed to the explicitly nonevolutionary accounts by antiteleologists such as Agassiz. That Darwin's arguments were most often aimed at the teleologists seems not to have been noticed by Neal C. Gillespie, who lumps together a host of different viewpoints under the rubric "special creation" (*Charles Darwin and the Problem of Creation*, pp. 22–3, 76–81).

67 B notebook, p. 84 (see also pp. 98–104, 115); C notebook, p. 184e; D notebook, p. 115.

68 B notebook, pp. 12–14.

69 Charles Lyell, *Principles of Geology*, 5th ed., 2:9–94 (Bk. III, chaps. 5–10). (During the period of the transmutation notebooks, Darwin used the 5th edition of Lyell's *Principles*.) Lyell's discussion of distribution often sounds as if all is a matter of chance fluctuations in inorganic, and hence organic, conditions, but this is not so. The hypothesis he was defending is fairly well stated in Book III, chapter 3: "Each species may have had its origin in a single pair, or individual, where an individual was sufficient, and species may have been created in succession at such times and in such places as to enable them to multiply and endure for an appointed period, and occupy an appointed space on the globe" (ibid., 2:55).

70 B notebook, p. 115: "It is a point of great interest to prove animals not adapted to each country – Provision for transportal otherwise not so numerous: quoted from Lyell."

71 Ibid., pp. 115, 130, 193–4; C notebook, p. 99.

72 See above, n. 43.

73 B notebook, pp. 46–47.

74 Ibid., p. 14.

75 Ibid., pp. 110–14. Darwin also commented on the Cuvier-Geoffroy dispute in his annotations of Whewell's *History of the Inductive Sciences* (see Edward Manier, *The Young Darwin and His Cultural Circle*, pp. 52–5).

76 B notebook, pp. 84, 99.

77 C notebook, pp. 76–7.

78 B notebook, p. 98.

79 Kohn, "Theories to Work By," pp. 79–86, 90–6, 113–40.

80 C notebook, p. 236.

81 E notebook, pp. 48–9.

82 See B notebook, pp. 101–2, 196.

83 Carpenter, *Physiology* (2nd ed., pp. 562–3; 1st ed., pp. 463–4).

84 Although Carpenter's book was published in 1839, reviewers noticed it, perhaps from proof sheets, in 1838.

85 D notebook, pp. 36–7. Silvan S. Schweber has recently discussed this passage, but without calling attention to its agreement with the commonly held view that the laws of physics and astronomy are parts of a purposeful system. I find it difficult to see how he reconciles this passage (and others, such as E notebook, pp. 48–49, quoted just above) with his statement that by the end of August 1838 Darwin was an agnostic ("The Origin of the *Origin* Revisited," pp. 255, 297). Compare Darwin's statement also with the quotation from Owen given in Chapter 1 p. 21.

86 B notebook, p. 210e: C notebook, p. 73; D notebook, pp. 36, 74e; E notebook, pp. 57, 71.

87 Georges Cuvier and A. Valenciennes, *Histoire Naturelle des Poissons*, p. 550.

88 Robert Boyle, *The Works of the Honourable Robert Boyle*, 5:413.

89 On this pre-*Origin* view of the economy of nature, see the following works by Camille Limoges: "Introduction," to C. Linné, *L'Equilibre de la Nature;*

Sélection Naturelle, pp. 59–69; "Economie de la Nature et la Idéologie Juridique chez Linné." See also Frank N. Egerton, "Changing Concepts of the Balance of Nature."

90 On the social, political, and religious interests served by natural religion, see M. C. Jacob, *The Newtonians and the English Revolution, 1689–1720;* John Redwood, *Reason, Ridicule and Religion,* esp. pp. 198–201; P. M. Heimann, "Science and the English Enlightenment." See also Basil Willey, *The Eighteenth Century Background,* esp. pp. 1-56. On the roots of Robert Boyle's philosophy of nature, see J. R. Jacob, "The Ideological Origins of Robert Boyle's Natural Philosophy."

91 On their spread to France and their influence on French biology, see Jacques Roger, *Les Sciences de la Vie dans la Pensée Française du XVIIIᵉ Siècle,* pp. 224–54. Without having looked deeply into the question, I would predict that among the proponents of holistic conceptions of life, nature, and society that were emerging, especially in Germany from the mid-eighteenth century, there would also have been a predilection for seeing in nature the same sorts of harmony, purpose, and adaptation.

92 Chapter 7 of Charles C. Gillispie, *Genesis and Geology,* usefully brings together quotations from several writers of the period. Paley and Malthus are discussed in my Chapter 3.

93 On Cuvier's science and the establishment in France, see Appel, "The Cuvier–Geoffroy Debate."

94 William Paley, *Natural Theology,* 1:112–14.

95 Limoges, *Sélection Naturelle,* pp. 69–70, 77–80. See also Camille Limoges, "Darwinisme et Adaptation," pp. 366–9.

96 I will suggest in Chapter 3 that Darwin's rejection of the harmonious view of nature occurred gradually after he read Malthus and that in some respects it was never complete.

97 Kohn, "Theories to Work By," pp. 98–9, 104–5, 143. See also W. Faye Cannon, "Charles Lyell, Radical Actualism, and Theory," pp. 118, 120 n. 28.

2. Darwin before Malthus

1 My second sense of perfect adaptation (see Chapter 1, pp. 33–5).

2 David Kohn, "Theories to Work By", pp. 83–96, 100–40.

3 See especially Sydney Smith, "The Origin of the *Origin*"; Gavin de Beer, "Introduction" to Darwin's E notebook, *Bulletin of the British Museum (Natural History),* Historical Series, 2 (1960):153–9; Camille Limoges, *La Sélection Naturelle.* Howard Gruber, *Darwin on Man;* Sandra Herbert, "The Place of Man in the Development of Darwin's Theory of Transmutation," Parts 1 and 2; Kohn, "Theories to Work By; Malcolm J. Kottler, "Charles Darwin's Biological Species Concept and Theory of Geographic Speciation."

4 De Beer, "Introduction" to Darwin's E notebook; Peter Vorzimmer, "Darwin, Malthus, and the Theory of Natural Selection," p. 533; Silvan S. Schweber, "The Origin of the *Origin* Revisited." Michael Ruse offers a

more subtle argument for the role of artificial slection in the origins of Darwin's theory: "Charles Darwin and Artificial Selection." My reading of the notebooks supports the conclusion of Limoges (*Sélection Naturelle*) and Kohn ("Theories to Work By") who deny that artificial selection played a significant role in leading Darwin to natural selection; Herbert has suggested this as well: "Darwin, Malthus, and Selection," pp. 211–13. It is worth noting that, as evidence for his claim that before Darwin read Malthus he was searching for an analogy between artificial and natural selection, Schweber cites only two of the several passages in which Darwin explicitly says that in nature offspring are *not* "picked" (Schweber, "*Origin* Revisited," pp. 235, 257).

5 B notebook, p. 1.

6 Ibid., pp. 2–4.

7 Ibid., pp. 16, 21. See Kohn: "the critical point is that he assumed heritable variability to be adaptive" ("Theories to Work By," p. 86).

8 C notebook, p. 236.

9 B notebook, p. 78; D notebook, p. 179 (D notebook pp. 152–79 were probably written September 11–28, 1838).

10 C notebook, p. 76.

11 D notebook, p. 166; see also p. 130.

12 B notebook, p. 219.

13 B notebook, p. 64 (my emphasis).

14 C notebook, p. 65; see also pp. 4, 84–5.

15 Limoges, *Sélection naturelle*, pp. 46, 76. See also Kohn, "Theories to Work By," p. 126.

16 B notebook, p. 227.

17 Robert C. Olby, *The Origins of Mendelism*, pp. 55–85; Peter Vorzimmer, *Charles Darwin*, pp. 30–42; Kottler, "Darwin's Biological Species Concept," pp. 288–91.

18 B notebook, p. 5; E notebook, p. 48. Cf. Kohn, "Theories to Work By," pp. 88–90, 105–7.

19 D notebook, p. 167.

20 B notebook, p. 210e: C notebook, p. 60.

21 B notebook, pp. 6–7.

22 R. B. Freeman and P. J. Gautrey, "Darwin's *Questions about the Breeding of Animals*, with a Note on *Queries about Expression*"; Peter Vorzimmer, "Darwin's Questions about the Breeding of Animals."

23 B notebook, pp. 24, 32–4, 120: "In intermarriages: smallest differences blended, rather stronger [have] tendency to imitate one of the parents; repugnance generally to marriage before domestication, afterwards none or little with fertile offspring: marriage never probably excepting from strict domestication, offspring not fertile or at least most rarely and perhaps never female – No offspring: physical impossibility to marriage." See also C notebook, p. 122.

24 B notebook, pp. 59, 75e, 120, 123e; C notebook, pp. 30, 51.

25 B notebook, pp. 24, 75e.

26 Ibid., p. 209e.

27 See the discussion in Kohn, "Theories to Work By," pp. 126–9.
28 B notebook, pp. 138, 140; the idea was fairly common. See for instance, Timothy Lenoir, "The Göttingen School in the Development of Transcendental Naturphilosophie in the Romantic Era."
29 D notebook, p. 13; also pp. 168–9 (an attempt to explain the law).
30 C notebook, pp. 34, 59, 66, 83–4.
31 Ibid., p. 149.
32 Ibid., p. 33 (but see pp. 51–2). Darwin applied the idea to the explanation of affinity and its distinction from analogy: ibid., pp. 138–40, 144, 202.
33 Ibid., p. 153.
34 Ibid., p. 136; see also pp. 2, 34.
35 D notebook, pp. 13–17.
36 Except hereditary monsters; ibid., pp. 14–15.
37 Ibid., p. 18.
38 B notebook, pp. 176–7e.
39 Ibid., p. 211.
40 Ibid., pp. 236, 239.
41 D notebook, pp. 174e–179.
42 Ibid., p. 175.
43 Ibid.
44 B notebook, pp. 224–7.
45 C notebook, pp. 51, 62–3, 124, 163, 171–3, 199.
46 See Richard W. Burkhardt, Jr., *The Spirit of System*, pp. 164–81. On Darwin's Lamarckism, see Herbert, "The Place of Man," Part 2, pp. 204–5; Kohn, "Theories to Work By," pp. 73, 156, n14, 129–33, 166–7, n118;" Robert J. Richards, "Influences of Sensationalist Tradition on Early Theories of the Evolution of Behavior."
47 C notebook, p. 174.
48 D notebook, pp. 174e, 177.
49 Burkhardt, *Spirit of System*, pp. 131–5.
50 DAR 42 (ser.3):MS p. 2v, in geological notes dated February 1835. Other portions of this manuscript refer explicitly to Lyell (See Chapter 1, p. 25). See Herbert, "Darwin, Malthus, and Selection," p. 215; "The Place of Man," Part 1, p. 233; Kohn, "Theories to Work By," pp. 70–2.
51 Dov Ospovat, "Lyell's Theory of Climate." On Lyell's uniformitarian metaphysics see Roy S. Porter, "Charles Lyell," pp. 395–433.
52 C notebook, pp. 146–7e.
53 B notebook, pp. 20–21, 36.
54 Burkhardt, *Spirit of System*, pp. 72–4, 131–5.
55 B notebook, p. 35; C notebook, pp. 167–9, 234. Darwin's difficulties in accounting for the extinction of the large animals were compounded by the fact that his geological investigations disclosed no change in circumstances significant enough to have resulted in their nonadaptation. This problem he only resolved after he formulated the theory of natural selection. The kind of struggle he thought was caused by Malthusian population pressure meant that even a very minor (but definite, rather

than oscillating) change could tip the balance in favor of one form or another. Compare *Journal of Researches* (London: Henry Colburn, 1839 [facs. reprint, New York: Hafner, 1952]), pp. 210–12, written in 1837, with the second (post-Malthus) edition: *The Voyage of the Beagle* (Garden City, N.Y.: Doubleday, 1962), pp. 174–7. Darwin's changing views on extinction in the pre-Malthus period are discussed at length in Kohn, "Theories to Work By," pp. 74–5, 156, n. 15, 77–9, 82, 96–100, 110–13, 116–17, and I have not thought it necessary to repeat here what he has said.

56 B notebook, p. 37.

57 Ibid., p. 170: see also p. 210e.

58 C notebook, p. 60.

59 Charles Lyell, *Principles of Geology*, 5th ed., 2:81–91 (Bk. III, chap. 10).

60 C notebook, p. 153.

61 D notebook, p. 69.

62 Georges Cuvier, *Le Règne Animal*, 1:18–19.

63 Lyell, *Principles* (5th ed. [American], 1837), 1:499, 507, 515, 527–8 (Bk. II chaps. 2–4). In saying this Lyell was merely giving expression to the view of variation that had been most common among naturalists since at least the mid-eighteenth century: except for Hybrids, most variations are caused by and correspond to changes in the external conditions under which a species lives. See Shirley Roe, "Rationalism and Embryology," pp. 30–7; Jacques Roger, *Les Sciences de la Vie dans la Pensée Française du XVIII* Siècle*, p. 568 (with reference to Buffon); Lenoir, "Transcendental Naturphilosophie"; Nils von Hofsten, "Linnaeus's Conception of Nature," p. 83; James L. Larson, *Reason and Experience*, p. 114; William Kirby, *On the Power, Wisdom, and Goodness of God*, . . . , 1:110–11.

64 It has recently been suggested that by July 1838 Darwin considered variations to be "indefinite, accidental, chance phenomena" and "not necessarily adaptive" (Schweber, "*Origin* Revisited," pp. 235–6, 264). No evidence is offered in support, nor have I found any supportive evidence in my reading of the notebooks.

65 In Darwin's pre-Malthus notebooks there is a conspicuous absence of any passage even remotely resembling Wallace's retort to Lyell: "Where is the balance?" (H. Lewis McKinney, *Wallace and Natural Selection*, p. 38).

66 A similar point was made as long ago as 1961 by Walter F. Cannon, who, however, had chiefly in mind ideas about nature that were held by only a portion of the scientific community, whereas the ideas I see as centrally important, such as perfect adaptation, were more widely shared. "The Bases of Darwin's Achievement." For a masterful extension of some aspects of Cannon's interpretation see James R. Moore, *The Post-Darwinian Controversies*, esp. pp. 307–51.

3. Natural selection and perfect adaptation, 1838–1844

1 D notebook, p. 135e.

2 On a superficial reading, one passage in the pre-Malthus notebooks

resembles the idea of natural selection (B notebook, p. 90). But on closer inspection it proves to include neither struggle, population pressure, nor the differential survival of variant forms. See David Kohn's perceptive analysis: "Theories to Work By," pp. 124–5; also pp. 102, 122, 146–8. In none of the pre-Malthus notebooks is there any more anticipation of natural selection than in the suggestion by Carpenter quoted above (Chapter 1, n. 49).

3 In addition to Kohn, just cited, see Peter Bowler, "Malthus, Darwin, and the Concept of Struggle"; Barry G. Gale, "Darwin and the Concept of a Struggle for Existence"; Sandra Herbert, "Darwin, Malthus, and Selection"; Camille Limoges, *La Sélection Naturelle*, pp. 79–81; Steven Shapin and Barry Barnes, "Darwin and Social Darwinism"; Peter Vorzimmer, "Darwin, Malthus, and the Theory of Natural Selection"; Robert Young, "Malthus and the Evolutionists"; Young, "The Historiographic and Ideological Contexts of the Nineteenth-Century Debate on Man's Place in Nature."

4 D notebook, pp. 134–35e. This passage appears to have been written in two sittings between September 28 and October 3, 1838. See Kohn, "Theories to Work By," pp. 140–1, 168–9, n. 132.

5 Thomas R. Malthus, *An Essay on the Principle of Population* (3rd ed.), 2:498–9.

6 See Howard E. Gruber, *Darwin on Man*, pp. 125–6. The Darwin notes Gruber refers to are in DAR 91:114–8. I am grateful to Nancy Mautner for sending me a copy.

7 William Kirby, *On the Power, Wisdom, and Goodness of God, . . .* , 1:157–9.

8 Malthus, *Essay* (3rd ed.), 2:501. That Malthus did not rule out all possibility of progress is insisted on by Samuel M. Levin, "Malthus and the Idea of Progress."

9 William Paley, *Natural Theology*, 2:155–6.

10 Ibid., 2:137–41.

11 Ibid., 2:139–49.

12 Malthus, *Essay on the Principle of Population* (1st ed.), pp. 116–30. See Daniel L. Le Mahieu, "Malthus and the Theology of Scarcity."

13 Malthus, *Essay* (3rd ed.), 2:306, 315–16 (in the 6th ed., which Darwin read [London: John Murray, 1826], this passage appears in vol. 2, p. 267).

14 The quoted phrase is from D notebook, p. 74e. Cf. Ernst Mayr, "Darwin and Natural Selection," p. 326: *"the struggle for existence is not a hopeless steady-state condition à la Malthus but the very means by which the harmony of the world is achieved and maintained."* Since I wrote this chapter James R. Moore's book has appeared, in which he calls attention to the relationship between Darwin's first statement of his theory and the theodicy of Malthus: *The Post-Darwinian Controversies*, p. 313.

15 These are some of the principal topics Darwin speculated on in his M and N notebooks (1838–9), which are transcribed and annotated by Paul H. Barrett and published in Gruber, *Darwin on Man*, pp. 266–305, 329–60. See esp. M notebook, pp. 27–32, 69–70.

16 C notebook, p. 166. See also ibid., pp. 171–3, quoted in Chapter 2, p. 52.

Elsewhere Darwin wrote, "By materialism, I mean, merely the intimate connection of kind of thought with form of brain"; in Darwin's copy of John Abercrombie, *Inquiries concerning the Intellectual Powers and the Investigation of Truth,* quoted in Edward Manier, *The Young Darwin and His Cultural Circle,* pp. 223–4.

17 For fuller discussion of this point see Dov Ospovat, "God and Natural Selection"; Moore, *Post-Darwinian Controversies,* pp. 318–22; W. Faye Cannon, "The Whewell-Darwin Controversy," p. 379; and Neal C. Gillespie, *Darwin and the Problem of Creation,* pp. 138–40; see also the discussion of nineteenth-century "materialism" in Maurice Mandelbaum, *History, Man, & Reason,* pp. 20–8.

18 Barrett, ed., M notebook, pp. 135–6, in Gruber, *Darwin on Man,* pp. 291–2 (Darwin filled these pages between September 8 and September 13, 1838). Pages 69–70 of the M notebook require a similar interpretation.

19 Another, later statement to the effect that superfecundity serves a useful purpose occurs in DAR 71:53–8. These reading notes on John Macculloch's *Illustrations and Proofs of the Attributes of God* (1837) are published in Gruber, *Darwin on Man,* pp. 416–22 (see p. 418).

20 Barrett, ed., M notebook, p. 154e, in Gruber, *Darwin on Man,* p. 296. See also E notebook, pp. 3, 6e; and DAR 71:53–8 (and in Gruber, *Darwin on Man,* p. 416).

21 E notebook, pp. 48–9 (written during the first week of November 1838). The passage is quoted in full in Chapter 1, p. 31.

22 DAR 71:53–8 (and in Gruber, *Darwin on Man,* pp. 416–19) (probably written in November 1838).

23 E notebook, pp. 4, 26e.

24 Ibid., pp. 50–1.

25 Cf. D notebook, p. 167 (discussed in Chapter 2, p. 49). At about this same time (November 7–December 2, 1838) Darwin wrote as if every variation were an adaptation: "Every structure is capable of innumerable variations, as long as each shall be *perfectly* adapted to circumstances of *times*" (E notebook, p. 57). There is ambiguity here, because it is at least possible that Darwin was referring only to the variations that are successful in the struggle for existence.

26 E notebook, pp. 111–12 (March 9–11, 1839); and Barrett, ed., N notebook, p. 42, in Gruber, *Darwin on Man,* p. 338. E notebook p. 58 is sometimes taken as an epitome of the theory of natural selection, but it is ambiguous on the crucial point of "accidental" variations.

27 DAR 71:58 (and in Gruber, *Darwin on Man,* p. 419). It should be observed that this is not a denial of the general idea of purpose in nature, any more than it was for Bacon or Owen.

28 At this point Darwin added, perhaps later: "& if so chance, that one out of every hundred litters is born with long legs."

29 DAR 205:5:28v (this portion of Darwin's notes on Macculloch is not published by Gruber and Barrett).

30 E notebook, pp. 68–9 (December 4–16, 1838). A superficially similar passage occurs in the B notebook, pp. 214–15, but there Darwin is saying

not that man is the result of "chance," but only that he could not be produced a second time, a supposition that accords well with the idea of a plan of creation. For an alternate interpretation, see Sandra Herbert, "The Place of Man in the Development of Darwin's Theory of Transmutation," p. 200.

31 D notebook, p. 135e (my emphasis). I was pleased to discover in an unpublished article sent me just as I was going to press with this volume that Jon Hodge has reached similar conclusions about Malthus's initial importance for Darwin's ideas on extinction and on the working out of natural selection *after* September 28 (M. J. S. Hodge, "Origins and Species I," p. 53.) As I suggested in Chapter 2, however, a substantial part of Darwin's "return to Lyell's position" on extinction occurred before he read Malthus.)

32 E notebook, p. 9e.

33 The shift from internal to external mechanisms is suggested in Frank N. Egerton, "Studies of Animal Populations from Lamarck to Darwin," pp. 246–7.

34 This may explain why early in November 1838 Darwin still did not think he knew the "laws of specific change" (E notebook, p. 54).

35 Charles Darwin, *Autobiography*, pp. 92–3.

36 "Essay of 1844," p. 254.

37 John C. Greene, "Reflections on the Progress of Darwin Studies," p. 246. See also Moore, *Post-Darwinian Controversies*, pp. 307–26.

38 See Gruber's remarks on the idea of plan as a competing theory; *Darwin on Man*, p. 211.

39 "Essay of 1844," pp. 133–4; "Sketch of 1842," p. 51.

40 *LLD*, 2:105. This Darwinian idea of design is discussed more fully in Ospovat, "God and Natural Selection." See also Chapter 9, pp. 225–6.

41 In Chapters 8 and 9 I will suggest that there are reasons for concluding that in some ways Darwin did not break with the traditional, essentially static conception of nature until the mid-1850s.

42 Most of the argument in this section is based on the "Essay of 1844," but reference is made to the "Sketch of 1842" when it seems that this may make Darwin's position clearer.

43 "Essay of 1844," pp. 91, 185, 196.

44 See Chapter 1, pp. 33–7.

45 DAR 71:58 (and in Gruber, *Darwin on Man*, p. 420; there are some errors in Barrett's transcription of this passage).

46 "Essay of 1844," p. 171; see also pp. 153–4 and "Sketch of 1842," p. 33n.

47 B notebook, p. 130. See also the works of Prichard and Agassiz cited in Chapter 1, nn. 42, 43.

48 "Essay of 1844," p. 153; *Origin*, p. 83.

49 DAR 71:53–8 (and in Gruber, *Darwin on Man*, p. 417).

50 E notebook, pp. 63, 71; see also p. 75.

51 "Essay of 1844," pp. 94–6.

52 Ibid., p. 85.

53 Ibid., pp. 95–6.
54 Even in the mature theory of the *Origin* this distinction is important. The differential adaptiveness of variations is an accident of circumstances and of the laws of variation, while relative adaptation is a product of natural selection. Cf. Robert N. Brandon, "Adaptation and Evolutionary Theory," where "relative adaptedness" is used to refer to the differential fitness of variations.
55 Kohn misleadingly suggests that differential adaptedness of variants and perfect adaptation of species are incompatible: "Theories to Work By," p. 105. See "Essay of 1844," p. 95, where at the top of the page Darwin assumes that there are differences in the adaptedness of variant individuals, while at the bottom he asserts that the species produced by natural selection are perfectly adapted.
56 As in Charles Darwin, *On the Various Contrivances by which British and Foreign Orchids Are Fertilised by Insects,* pp. iii, 1–2, 24.
57 "Essay of 1844," pp. 57–9, 78.
58 Darwin's formulation of this proposition may have owed something to his reading of J. F. Blumenbach, *An Essay on Generation:* "There is nothing unreasonable in supposing, that after the subjection of these animals, their whole frame and oeconomy suffer a very great change; that then the Formative Nisus deviates in some degree from its original laws, and therefore these animals as they degenerate into endless varieties, are also more subjected to monstrosity" (pp. 83–4).
59 "Essay of 1844," pp. 77–8.
60 Ibid., p. 83 (Darwin's emphasis).
61 Ibid., pp. 84–5; "Sketch of 1842," pp. 4–5.
62 "Essay of 1844," pp. 82–4.
63 *Origin,* p. 45.
64 "Essay of 1844," p. 83.
65 Ibid., pp. 84–5.
66 Compare p. 83 with pp. 84–5 in the "Essay of 1844."
67 Ibid., p. 85.
68 This is probably partly responsible for Darwin's greater emphasis on the role of sports, of rapid change, and so forth, in the "Essay" than in the *Origin*. See Francis Darwin's "Introduction" to *Foundations of the Origin of Species,* p. xxix. For an interesting treatment of Darwin's views on variation, which does not, however, take into account the distinction I have just discussed, see Peter Bowler, "Darwin's Concepts of Variation."
69 "Essay of 1844," p. 95.
70 Ibid., pp. 59, 81, 83; "Sketch of 1842," p. 4.
71 "Sketch of 1842," pp. 1–2, 4–5, 8–9, 13–15 (Darwin treated crossing as an indirect effect of a change of conditions), 20, 21, 30–1, 32–3, 34n, 37; 'ESSAY OF 1844," pp. 57–76, 83, 84–5, 90–2, 110, 120, 145, 153–4, 183–94.
72 "Sketch of 1842," p. 53.
73 "Essay of 1844," pp. 90–1.
74 E notebook, p. 71 (mid-December 1838).
75 Ibid., p. 122e.

76 DAR 16.2:303.
77 C notebook, pp. 174–5.
78 *Origin*, p. 201.
79 Ibid., p. 113; see also *LLD*, 1:480.
80 "Sketch of 1842," p. 5; "Essay of 1844," p. 85.
81 *Origin*, p. 45.
82 Ibid. Francis Darwin long ago noted this difference: *Foundations*, pp. xxviii–xxix, 81n.
83 Cf. Gruber, *Darwin on Man*, pp. 56, 124–7. J. Maynard Smith has remarked that "if species were perfectly adapted evolution would cease" ("Optimization Theory of Evolution," p. 38). Until the 1850s this was Darwin's view, but he thought it ceased not because selective pressure keeps the perfect form constant, but rather because when forms are perfectly adapted there are no variations on which selection can work.
84 *Origin*, p. 84.
85 "Essay of 1844," p. 109.

4. Part II of Darwin's work on species

1 C notebook, pp. 76–7, 177; D notebook, p. 69.
2 For one explanation of why "difficulties" are included in Part I, see M. J. S. Hodge's interesting discussion of the structure of Darwin's argument in the essays of the 1840s, *Natural Selection*, the *Origin*, and Darwin's projected multivolume work: "The Structure and Strategy of Darwin's 'Long Argument.'"
3 Charles Darwin, *Autobiography*, p. 118. The "Black Box" material, much of which was dated at the time it was written, makes it clear that it was in 1839, when the E notebook was concluded, that Darwin began his practice of keeping most of his notes "in bulk" (*MLD*, 1:76; *LLD*, 1:128–9).
4 These envelopes have been assigned the following numbers: DAR 205.1 (abortive organs), DAR 205.3 and DAR 205.4 (geographical distribution). DAR 205.5 (classification/divergence), DAR 205.6 (embryology), DAR 205.9 (paleontology and extinction). Each envelope probably contains the material from one of the portfolios mentioned in the *Autobiography* (pp. 137–8). I am indebted to Sydney Smith for helpful information on Darwin's filing system and more generally for sharing with me his unparalleled knowledge of Darwin's habits of work.
5 Charles Darwin, *The Variation of Animals and Plants under Domestication*, 1:13–21.
6 *Athenaeum*, no. 1854 (May 9, 1863), p. 617; *LLD*, 2:207.
7 *LLD*, 2.163–4. Some years later, in a letter to H. N. Moseley, Darwin wrote: "As long as a man believes in evolution biology will progress, and it signifies comparatively little whether he admits natural selection and then gains some light on the method, or remains in utter darkness" (DAR 146). This letter was brought to my attention by Stan Rachootin.
8 Huxley MSS, vol. 5, fol. 252, Darwin to T. H. Huxley, January 22 [1862], Archives, Imperial College, London. On the importance of the Part II

chapters see also *LLD*, 1:519–23; *LLD*, 2:6–7, 57–8, 130–3, 161–2; *MLD*, 1:144–5, 172–4.

9 "Sketch of 1842," p. 7; *LLD*, 1:473.

10 Darwin, *Autobiography*, p. 124.

11 Ibid., p. 119: "I worked on true Baconian principles, and without any theory collected facts on a wholesale scale, more especially with respect to domesticated productions, by printed enquiries, by conversation with skilful breeders and gardeners, and by extensive reading" (see also *LLD*, 1:411–12). On the role of the breeders, see Chapter 2, n. 4.

12 *Natural Selection*, bibliography (pp. 587–629).

13 Sandra Herbert, "The Place of Man," Part I, pp. 240–5; see also Malcolm J. Kottler, "Charles Darwin's Biological Species Concept and Theory of Geographic Speciation," p. 281.

14 Sandra Herbert, "The Place of Man," Part II, pp. 156–7.

15 DAR 73:51a; DAR 205.1; *MLD*, 1:59–61, 415; Rev. Richard Owen, *The Life of Richard Owen*, 1:108, 188, 209, 292.

16 DAR 205.9 (letters), dated November 11, 1854.

17 DAR 181.

18 Darwin, *Autobiography*, p. 115. Contrast *LLD*, 1:288.

19 *LLD*, 1:387–8.

20 This is Darwin's date, but it appears that he went on adding questions to the list for use at later meetings (DAR 206 [inserted in Questions & Experiments notebook]).

21 See Frank N. Egerton, "Darwin's Method or Methods?" pp. 285–6.

22 See B notebook, pp. 227–8.

23 Darwin, *Autobiography*, pp. 119, 137–8.

24 Ibid., p. 119; *LLD*, 1:384.

25 Richard Owen, "Description of the Cranium of the *Toxodon platensis*."

26 Charles Darwin, *Voyage of the Beagle*, p. 83.

27 *Origin*, pp. 329–33. On Owen's law see Chapter 5, pp. 136–8.

28 *Origin*, p. 456; *LLD*, 2:131.

29 See Chapter 1, pp. 19–20.

30 Darwin's Library, Alphonse De Candolle, *Géographie Botanique Raisonée*, 1:463.

31 DAR 113.

32 Darwin's Library, De Candolle, *Géographie Botanique Raisonée*, 1:588. I am indebted to Janet Browne for calling my attention to Darwin's extensive annotations in this book.

33 Darwin's Library, Henri Milne Edwards, *Introduction à la Zoologie Générale*, p. 154.

34 "Essay of 1844," p. 133.

35 *Origin*, p. 312.

36 "Essay of 1844," pp. 133, 252–3; see also *Origin*, pp. 438–9.

37 B notebook, p. 104; *LLD*, 2:80, 155.

38 "Essay of 1844," pp. 134–6.

39 *Origin*, pp. 312, 416–18, 434–5, 439–43.

40 The best instance of Darwin's testing a hypothesis is perhaps his collecting

botanical statistics to support his views on classification and divergence (see Chapter 7; see also Janet Browne, "Darwin's Botanical Arithmetic and the 'Principle of Divergence,' 1854–1858").

41 It would perhaps be better to say that this was a consequence of his perspective as a transmutationist *and* of his tendency to think of natural laws as governing rules or causative agents.

42 DAR 205.5:40, dated February 1841.

43 William Sharp MacLeay, *Horae Entomologicae*. MacLeay's system is discussed in Mary P. Winsor, *Starfish, Jellyfish and the Order of Life*, pp. 81–6.

44 W. S. MacLeay, "Remarks on the Identity of Certain General Laws which Have Been Lately Observed to Regulate the Natural Distribution of Insects and Fungi," (hereinafter "Insects and Fungi"), pp. 46–7.

45 MacLeay, *Horae Entomologicae*, pp. 88, 169–70; W. S. MacLeay, "Annulosa," p. 8.

46 MacLeay, "Annulosa," p. 8n.

47 MacLeay, "Insects and Fungi," p. 48.

48 MacLeay, *Horae Entomologicae*, pp. 328–9.

49 Ibid., p. 330.

50 J. B. Lamarck, *Histoire Naturelle des Animaux sans Vertèbres*, p. 457; MacLeay, *Horae Entomologicae*, p. 332.

51 MacLeay, *Horae Entomologicae*, p. 170.

52 MacLeay admitted that he knew no botany and was only guessing that the scheme applied among plants, on the grounds that nature would not follow one plan for animals and another for plants.

53 The names of categories were unimportant, MacLeay said. Later he elevated his classes to the status of subkingdoms, orders to classes, etc. ("Annulosa," pp. 7–9).

54 MacLeay expressed doubts as to whether *Ornithorhynchus*, the most notorious aberrant form, was truly osculant (*Horae Entomologicae*, pp. 266–7). On fewness of forms in osculant groups see ibid., p. 37.

55 Ibid., p. 170.

56 Ibid., pp. 263–4, 272.

57 Ibid., p. 401.

58 Ibid., p. 363.

59 See, e.g., the comments of Henri Milne Edwards, "Considérations sur Quelques Principes Relatifs à la Classification Naturelle des Animaux," pp. 78–9; and Richard Owen, "Address," p. lxvii.

60 W. S. MacLeay, "A Reply to some Observations of M. Virey in the 'Bulletin des Sciences Naturelles, 1825,'" p. 49. On this point I am indebted to a perceptive analysis by Camille Limoges and Jean-Claude Cadieux (unpublished).

61 Richard Owen, "On the Anatomy of the Dugong," pp. 39–40.

62 Richard Owen, *Lectures on the Comparative Anatomy and Physiology of the Invertebrate Animals*, glossary.

63 MacLeay, *Horae Entomologicae*, p. 366; "Annulosa," p. 7.

64 MacLeay, "Insects and Fungi," pp. 52–3n.

65 MacLeay, *Horae Entomologicae*, pp. 163–5, 324, 362–3.

66 Ibid., p. 163.

67 MacLeay, "Insects and Fungi," p. 63.

68 The quoted phrase is from C. F. Schonbein, as given in Charles C. Gillispie, *Genesis and Geology*, p. 200.

69 G. R. Waterhouse, "Descriptions of Some New Species of Exotic Insects."

70 Richard Owen Collection, Notebooks 3 and 9, British Museum (Natural History); Owen MSS, 276.h² 27 and Hunterian Lectures for 1837, Royal College of Surgeons.

71 J. E. Gray, "Arrangement of Reptiles."

72 [Robert Chambers], *Vestiges of the Natural History of Creation*, reprint of the 1st ed., pp. 191, 232, 236 ff.

73 See Winsor, *Starfish*, pp. 87–97.

74 *LLD*, 1:252: B notebook, p. 179. On Darwin's reading of *Horae Entomologicae*, see Sydney Smith, "The Darwin Collection at Cambridge with One Example of Its Use," p. 100.

75 E.g., Gavin de Beer, "Introduction" to "Darwin's Notebooks on Transmutation of Species," p. 29; Howard E. Gruber, *Darwin on Man*, p. 196. David Kohn has suggested a more positive role for MacLeay's ideas in Darwin's intellectual development ("Theories to Work By"; pp. 76, 157, n. 16).

76 E.g., B notebook, p. 28: "?How is it that there come aberrant species in each genus (with well characterized parts belonging to each) approaching another."

77 B notebook, p. 27: see also C notebook, pp. 139, 150.

78 Darwin to G. R. Waterhouse, [December 1843], British Museum (Natural History).

79 D notebook, pp. 58–9.

80 B notebook, p. 28; C notebook, pp. 201–2; DAR 205.5:59 (undated, approximately 1840); DAR 205.5:60, dated April 1843.

81 This is implied in early uses of the tree image (B notebook, pp. 19–28); it is stated explicitly in C notebook, p. 155.

82 DAR 185, Darwin to G. R. Waterhouse, postmarked July 27, 1843.

83 B notebook, pp. 23–4. Darwin's use of "affinity" here may be a slip of the pen for "analogy," or it may indicate that he had not yet mastered the distinction or learned to use the terms consistently.

84 B notebook, pp. 45–6, 263. It is the suggestion that the lack of an atmospheric type would leave four groups that makes it clear that Darwin thought his "law" would normally produce groups of five (the parenthetical 7 refers to Edward Newman's substitution of seven circles for MacLeay's five: De Beer, "Introduction" to B notebook, p. 29).

85 C notebook, p. 73; B notebook, p. 57; D notebook, p. 62e. Darwin's initial efforts to account for the number five belie the suggestion that from the outset he saw his irregularly branching tree of life as incompatible with the regularity of MacLeay's groups (cf. Gruber, *Darwin on Man*, p. 196).

86 B notebook, p. 162 (Darwin attributed the remark to Waterhouse). See also B notebook, p. 129: "The relation of analogy of MacLeay etc. appears to me the same as the irregularities in the degradation of structure of Lamarck, which he says depends on external influences."

87 B notebook, pp. 55e–56e; C notebook, pp. 62–3, 73–4, 100, 111e–16e.

88 C notebook, pp. 139–42e; DAR 205.5:38 and 44, two undated fragments (approx. 1839–42). Darwin drew on his law of hereditary fixing of characters to explain how as a practical matter one might distinguish between affinity and analogy (D notebook, p. 51). See Kohn, "Theories to Work By," pp. 165–6, n. 107–8. On the application of the idea of hereditary fixing to classification generally, see Chapter 2, pp. 46–7.

89 B notebook, p. 58: "I cannot understand the universality of such law."

90 Similarity of conditions was Darwin's principal, but not only, explanation of analogies. He also suggested "laws of variation" as a cause, such as reversion and inheritance of a tendency to vary in similar ways (B notebook, p. 58; DAR 73:37; *Origin*, p. 159).

91 *Origin*, p. 428.

92 See David E. Allen, *The Naturalist in Britain*, p. 102, on MacLeay and his theory: "Its proponent was a man of no small academic stature and the very extensive data put forward in its support demanded that it be taken seriously by scholars."

93 Reported in the *Athenaeum*, no. 829 (September 16, 1843), p. 851. Darwin's notes on the *Athenaeum* report are in DAR 205.5. Strickland, Edward Blyth, and others had previously rejected quinarianism (Blyth, "Observations on the Various Seasonal and Other External Changes which Regularly Take Place in Birds . . ."; Peter Rylands, "On the Quinary, or Natural, System of M'Leay, Swainson, Vigors, &c."; Strickland, "Observations upon the Affinities and Analogies of Organized Beings."

94 DAR 205.5:96, dated March 1844.

95 DAR 205.5:108, dated November 1844. This makes highly improbable the argument presented in Gruber, *Darwin on Man*, p. 196n, that Darwin undertook his 1846–54 systematic work on barnacles in order to show the inadequacies of MacLeay's system.

96 De Beer, "Introduction to Darwin's Notebooks," p. 27.

5. Natural history after Cuvier

1 That there was a post-Cuvier compromise weighted toward Geoffroy's position was proposed independently by Toby Appel and me (Toby Appel, "The Cuvier-Geoffroy Debate and the Structure of Nineteenth Century French Zoology"; Dov Ospovat, "Embryos, Archetypes, and Fossils"). To some extent we were both anticipated by Patrick Geddes in the article "Morphology," *Encyclopaedia Britannica*, 9th ed. See also William Coleman, "Morphology between Type Concept and Descent Theory," p. 160. E. S. Russell misleadingly described midcentury morphology as a return to the tradition of Cuvier (*Form and Function*, p. 195).

2 It should be noted, however, that although pre-*Origin* morphology did not require the idea of descent, neither did it exclude it, as Coleman seems to say: ("Morphology between Type Concept and Descent Theory," p. 150).

3 It is important that this new developmental view of nature not be confused with the "development hypothesis," which was a term common-

ly used to describe theories of transmutation, such as those of Chambers and Lamarck. Some transmutationists had as little sense for morphological or historical *development* as Bonnet displayed in his scale of nature; and on the other hand the only *real* development the majority of morphologists admitted was the observable development of the embryo.

4 Wallace and Chambers both drew on paleontological laws worked out by Owen and others, and Spencer found the inspiration for his view of progressive evolution in the lectures and writings of Owen, Carpenter, and von Baer. Alfred Russel Wallace, "On the Law which has Regulated the Introduction of New Species"; "On the Tendency of Varieties to Depart Indefinitely from the Original Type," pp. 268-79; [Robert Chambers], *Vestiges of the Natural History of Creation*, p. 212; *Vestiges*, 10th ed., pp. 147-54; Herbert Spencer, "Progress," pp. 9-10; Spencer, "The Filiation of Ideas," pp. 541-2.

5 Very general representations of archetypes were often employed also by those who favored a linear conception of nature – several of the *Naturphilosophen*, for instance. On the notion of ideal types as it developed in late-eighteenth-century Germany, see Timothy Lenoir, "Generational Factors in the Origin of *Romantische Naturphilosophie*."

6 The work of Victor Audouin on articulates is an exception ("Recherches Anatomiques sur le Thorax des Animaux Articulés et Celui des Insects Hexapodes en Particulier"). Audouin is of the same generation as von Baer, Owen, and Milne Edwards and might well have been included in the present discussion. Geoffroy, for all his contributions to morphology, seems to have had no very clear conception of an archetype. On early-nineteenth-century morphology see Russell, *Form and Function*, pp. 52-101.

7 Karl Ernst von Baer, *Ueber Entwickelungsgeschichte der Thiere*, vol. 1, plate 3 (reproduced in Russell, *Form and Function*, p. 119).

8 Von Baer, *Entwickelungsgeschichte*, 1:199-262; see also "Ueber das äussere und innere Skelet"; "Beiträge zur Kenntniss der Niedern Thiere"; and *Entwickelungsgeschichte*, 2:57-102.

9 On recapitulation, see Stephen J. Gould, *Ontogeny and Phylogeny*, pp. 33-52.

10 For an interesting exception, see Toby Appel, "Henri de Blainville and the Animal Series."

11 Lorenz Oken, *Elements of Physiophilosophy*, pp. 491-4. See Gould, *Ontogeny and Phylogeny*, pp. 39-45.

12 Von Baer, *Entwickelungsgeschichte*, 1:207-8. I have generally followed Huxley's translation of the "Fifth Scholium," which I will cite as Huxley, trans., "From the Works of von Baer." The reference above is to pp. 195-7.

13 Ibid., p. 220 (Huxley, trans., "From the Works of von Baer," pp. 209-10).

14 Ibid., pp. 221-5 (Huxley, trans., "From the Works of von Baer," pp. 211-14).

15 Ibid., pp. 230-1 (Huxley, trans., "From the Works of von Baer," pp. 219-20).

16 Russell, *Form and Function*, p. 129.
17 Georges Cuvier, *Le Règne Animal*, 1:xx–xxi. See William Coleman, *Georges Cuvier, Zoologist*, p. 149.
18 Georges Cuvier and A. Valenciennes, *Histoire Naturelle des Poissons*, 1:568–9. See Coleman, *Cuvier*, p. 153.
19 Von Baer, *Entwickelungsgeschichte*, 1:236 (Huxley, trans., "From the Works of von Baer," p. 225).
20 On the spread of von Baer's views see Dov Ospovat, "The Influence of Karl Ernst von Baer's Embryology, 1828–1859."
21 Von Baer, *Entwickelungsgeschichte*, 1:225–9, 237–42 (Huxley, trans., "From the Works of von Baer," pp. 215–18, 226–31); and "Beiträge zur Kenntniss der Niedern Thiere," pp. 739–59.
22 Jane Oppenheimer called attention to this in "An Embryological Enigma in the *Origin of Species*."
23 Martin Barry, "On the Unity of Structure in the Animal Kingdom," p. 118.
24 Ibid., pp. 347–8.
25 Ibid., pp. 136–7, 362–3.
26 Henri Milne Edwards, "Observations sur les Changemens de Forme que Divers Crustacés Eprouvent dans le Jeune Age."
27 Henri Milne Edwards, "Considérations sur Quelques Principes relatifs a la Classification Naturelle des Animaux," p. 66.
28 Ibid., pp. 66–9.
29 Ibid., pp. 69–76.
30 DAR 72:117.
31 Milne Edwards, "Classification Naturelle," p. 70
32 This feature of Owen's work was noted by Russell, *Form and Function*, pp. 102–3.
33 Owen Collection, *Notes and Synopses of Lectures, 1828–41*, British Museum (Natural History).
34 Owen MSS, Hunterian Lectures for 1837, Royal College of Surgeons.
35 Richard Owen, *Lectures on the Comparative Anatomy and Physiology of the Invertebrate Animals*, pp. 367–71.
36 He argued cogently that it had limited value as a guide to the discovery of homologies (*On the Archetype and Homologies of the Vertebrate Skeleton*, pp. 89, 104–6).
37 Ibid., p. 73. Owen's *Archetype* first appeared in the *Report of the British Association for the Advancement of Science, 1846*.
38 Owen, *Archetype*, pp. 81, 136.
39 Ibid., pp. 81, 102–3.
40 Ibid., pp. 171–2.
41 Ibid., plate II and pp. 176–7 (Owen's figure of the archetype is also given in Russell, *Form and Function*, p. 105).
42 Owen, *Archetype*, p. 136.
43 Richard Owen, "Description of the Cranium of the *Toxodon Platensis*"; *Description of the Skeleton of an Extinct Gigantic Sloth (Mylodon robustus, Owen)*, p. 163; "Fossil Mammalia," p. 55.

44 Richard Owen, "Description of Teeth and portions of Jaws of Two Extinct Anthracotheroid Quadrupeds . . ."

45 William Buckland, *Geology and Mineralogy considered with reference to Natural Theology,* 1:114; Louis Agassiz, "On the Differences between Progressive, Embryonic, and Prophetic Types"; W. S. MacLeay, *Horae Entomologicae,* pp. 266-7.

46 William Coleman, "Limits of the Recapitulation Theory."

47 Carl Vogt, *Embryologie des Salmones,* pp. 256-61.

48 Or, among invertebrates, within the separate classes.

49 Louis Agassiz, *Monographie des Poissons Fossiles du Vieux Grès Rouge ou Systeme Dévonien (Old Red Sandstone),* p. xxvi.

50 Richard Owen, "Report on British Fossil Reptiles, Part II," p. 201; *Lectures on Invertebrates,* p. 129.

51 Owen Collection, *Manuscripts, Notes and Synopses of Lectures.* 1849-64, Hunterian Lectures for 1853, British Museum (Natural History).

52 William B. Carpenter, *Principles of Physiology, General and Comparative,* pp. viii-ix; [Richard Owen], "Lyell – on Life and its Successive Development," p. 430; [Richard Owen], "Generalizations of Comparative Anatomy," p. 48.

53 [Owen], "Lyell – on Life and its Successive Development," pp. 448-50.

54 [Owen], "Generalizations of Comparative Anatomy," pp. 51-2.

55 Owen himself took it to be evidence of descent, though not of transmutation. For his distinction between these ideas, see [Richard Owen], "Darwin on the Origin of Species," pp. 501-3. John H. Brooke, in discussing Owen's views, seems not to have considered the possibility of such a distinction ("Richard Owen, William Whewell, and the *Vestiges*").

56 Thomas H. Huxley, "The Coming of Age of 'The Origin of Species'" (1880), p. 241.

57 See Martin J. S. Rudwick, *The Meaning of Fossils,* pp. 239-67; and Daniel P. Todes, "V. O. Kovalevskii: The Genesis, Content, and Reception of His Paleontological Work."

58 Thomas H. Huxley, "On Certain Zoological Arguments Commonly Adduced in Favour of the Hypothesis of the Progressive Development of Animal Life in Time" (1855), in *The Scientific Memoirs of Thomas Henry Huxley,* 1:300-4; Huxley, "On the Persistent Types of Animal Life" (1859), in *Scientific Memoirs,* 2:90-3. Huxley remained a nonprogressionist until 1868 ("On the Animals which are Most Nearly Intermediate between Birds and Reptiles" [1868], in *Scientific Memoirs,* 3:303-13). See Michael Bartholomew, "Huxley's Defence of Darwin."

59 Thomas H. Huxley, *American Addresses,* pp. 83-84, 90. Only after Huxley said that this European series of fossils was proof of evolution did he go on to discuss the even more impressive American series then just discovered. For his European series Huxley used *Plagiolophus* (Cuvier's *Palaeotherium minus*) from the Eocene, *Anchitherium* (Cuvier's *Palaeotherium aurelianense*) from the upper Eocene and lower Miocene, *Hipparion* from the Miocene, and a species of *Equus* from the latter part of the Miocene. Owen in 1851 used two unspecified species of *Paleotherium,* one each

from the Eocene and Miocene; *Hipparion,* and a species of *Equus* from the Pliocene. Owen also discussed these same fossils in his lectures at the School of Mines in 1857 and again, this time as evidence of descent, in 1868 in his *On the Anatomy of Vertebrates,* 3:791–2, 825. It was in the following year that Huxley "happened to be looking critically into the bearing of paleontological facts upon the doctrine of evolution" and discovered that the series of horselike fossils provided "demonstrative evidence" of it (*American Addresses,* pp. 83–4).

60 Owen, *Anatomy of Vertebrates,* 3:789–92; the references to Cuvier are to *Recherches sur les Ossemens Fossiles,* 1:lvii.

61 And also because of its dissemination through Spencer's evolutionary writings; see above, n. 4.

62 Richard Owen, *On the Nature of Limbs,* p. 84.

63 I have in mind particularly Owen's delicate position as an employee of the Royal College of Surgeons, to which his letters in the Owen MSS, Royal College of Surgeons, and Owen Collection, British Museum (Natural History) give ample testimony.

64 Richard Owen, "Teleology of the Skeleton of Fishes"; *Limbs,* p. 86.

65 See Chapter 1, p. 21.

66 See the perceptive discussion by John H. Brooke, "Natural Theology and the Plurality of Worlds."

67 R. S. S. Baden-Powell, *Essays on the Spirit of the Inductive Philosophy,* . . . pp. 108, 135, 253–8, 261–2, 400–1, among others.

68 Brooke, "Brewster-Whewell Debate," pp. 232–33.

69 [Whewell], *Plurality of Worlds;* pp. 240–7, 253, 270, 274, 371. This of course was an old argument; see Robert Boyle, *The Works of the Honourable Robert Boyle,* 5:423.

70 [Whewell], *Plurality of Worlds,* pp. 246–7. In discussing the significance of Whewell's position John Brooke, it seems to me, does not consider the possibility that Whewell distinguished between a universal law of development and a law of development established at the time of the creation of life on earth. If he was making such a distinction, it would not be nonsensical to see his essay as anti-*Vestiges* but at the same time, in allowing the possibility of transmutation, "proto-Darwinian" ("Brewster-Whewell Debate," p. 272).

71 See for instance J. Piveteau, "Vertebrate Paleontology," p. 447; Todes, "Kovalevskii," pp. 120–3, 141. Peter Bowler's much more satisfactory account stresses the significance of the new interpretations of the fossil record before 1859 (*Fossils and Progress,* esp. pp. 93–115).

72 See especially Owsei Temkin, "German Concepts of Ontogeny and History around 1800"; see also Gould, *Ontogeny and Phylogeny,* pp. 33–63; and Lenoir, "Generational Factors." On the transition to a historical viewpoint in geology, see W. R. Albury and D. R. Oldroyd, "From Renaissance Mineral Studies to Historical Geology, in the Light of Michel Foucault's *The Order of Things.*"

73 This also was not an isolated development. See Michel Foucault, *The Order of Things,* pp. 226–32, 263–79.

74 See the discussion in Maurice Mandelbaum, *History, Man, & Reason*, pp. 41–9.

75 Von Baer, *Entwickelungsgeschichte*, 1:263–4; "Das Allgemeinste Gesetz der Natur in aller Entwickelung," 1:71.

76 Owen, *Limbs*, pp. 84–6; Milne Edwards, *Introduction à la Zoologie Générale*, pp. v–vi.

77 See the suggestive remarks of Temkin, "German Concepts of Ontogeny and History," p. 246. Though Owen did not openly advocate descent before 1859, his views led others to suppose he believed in it. See, e.g., Leonard G. Wilson, ed., *Sir Charles Lyell's Scientific Journals on the Species Question*, p. 241.

6. Darwin and the branching conception

1 E. S. Russell, *Form and Function*, p. 247.

2 Ernst Mayr, *Animal Species and Evolution*, pp. 595–601; David L. Hull, "The Metaphysics of Evolution," p. 325.

3 "On my theory an '*Exemplar*' is no more wanted than to account for the likeness of members of one Family" (DAR 205.5:143, notes on William Whewell's *Of the Plurality of Worlds* [read by Darwin in May 1854]).

4 Darwin's Library, Richard Owen, *On the Nature of Limbs*. Darwin subsequently added: "I follow him that there is a created archetype, the parent of its class."

5 Cf. Russell, *Form and Function*, pp. 216, 230, 232.

6 B notebook, p. 206; C notebook, pp. 51–2; E notebook, p. 89.

7 DAR 74:112v. See also *MLD*, 1:415; and DAR 205.1:47, dated March 1846.

8 *Origin*, pp. 438–9.

9 Charles Darwin, *A Monograph on the Sub-Class Cirripedia*, 2:34–40; *A Monograph on the Fossil Lepadidae*, p. 48. On Darwin's cirripede work see Mary P. Winsor, "Barnacle Larvae in the Nineteenth Century," pp. 305–7; see also Michael T. Ghiselin and Linda Jaffe, "Phylogenetic Classification in Darwin's *Monograph on the Sub-Class Cirripedia*."

10 Russell, perceptive as usual, remarked long ago that the evidence from morphology probably contributed greatly to the success of Darwin's theory (*Form and Function*, p. 238). See also George Gaylord Simpson, "Anatomy and Morphology."

11 Asa Gray, *Darwiniana*, p. 237. On the background to Gray's remark see James R. Moore, *The Post-Darwinian Controversies*, pp. 269–80.

12 *LLD*, 2:367.

13 My theory "unites these two grand classes of views," he wrote in March 1847 (DAR 72:161). It is important always to remember that for the naturalists of Darwin's generation "teleology" usually refers not to the mere idea of purpose, but rather to a particular approach to biological explanation.

14 DAR 205.5:143.

15 *Origin*, p. 206.

16 To Cuvier's claim that the principle of unity of composition is "quite subordinate 'à celui des conditions d'existence'" Darwin responded "I dispute this" (DAR 74:38). See also B notebook, pp. 112–13; and Camille Limoges, "L'Economie Naturelle et le Principe de Corrélation chez Cuvier et Darwin."

17 *Origin*, p. 206; *Natural Selection*, pp. 383–4.

18 It was perhaps prompted by the recent dispute between Huxley and Hugh Falconer on the Cuvierian method in paleontology. See Houston Peterson, *Huxley*, pp. 84–8; and *MLD*, 1:89–90. Darwin's comment to Hooker is worth noting, that although he inclined to Huxley's (anti-Cuvierian) side, "to deny all reasoning from adaptation and so-called final causes, seems to me preposterous" (DAR 114:170).

19 DAR 74:112.

20 Charles Darwin, *The Descent of Man and Selection in Relation to Sex*, p. 61 (Part I, chapter 2); Darwin's Library, Owen, *On the Nature of Limbs*, index sheet pinned in back.

21 For fuller discussion on this point see Dov Ospovat, "God and Natural Selection," and Maurice Mandelbaum, *History, Man & Reason*, pp. 85–7.

22 *LLD*, 2:367.

23 DAR 71:45 (referring to Owen, *On the Nature of Limbs*, pp. 13, 14, 39). A. J. Cain has made some useful comments on Darwin and Owen, but they must be read with care, for it does not seem to me that he has rightly appreciated what Owen and Darwin were saying or recognized the antievolutionary implications of the strict teleology against which they were arguing ("The Perfection of Animals," pp. 36–47).

24 *Natural Selection*, pp. 379–84.

25 DAR 71:45.

26 B notebook, p. 1.

27 B notebook, p. 78; D notebook, pp. 174e, 179.

28 E notebook, pp. 83–4.

29 B notebook, pp. 111, 163; C notebook, p. 48e. The reference is to Richard Owen, "Remarks on the *Entozoa*."

30 Darwin's Library, Johannes Müller, *Elements of Physiology*, 1:398–9; 2:1591–2.

31 "Essay of 1844," p. 219; "Sketch of 1842," p. 42. On Darwin and re-capitulation see Dov Ospovat, "The Influence of Karl Ernst von Baer's Embryology, 1828–1859," pp. 24–7; and Stephen J. Gould, *Ontogeny and Phylogeny*, pp. 70–4.

32 *Origin*, p. 449.

33 DAR 205.6:11, dated January 1840 ((Darwin's additions)).

34 "As selection does not act on young state, the variation with tendency to be hereditary, at that age, must be counteracted by crossing – but not so with such variation in older age, for here selection acts – Hence young or foetus in wild species will generally agree closer than old form" (DAR 205.6:17, dated February 1841).

35 Ibid.

36 DAR 205.6:19 (undated). See "Sketch of 1842," p. 43.

37 "Essay of 1844," pp. 221–5.
38 "Sketch of 1842," p. 44n.
39 DAR 205.6:27 (undated); *Origin*, p. 444.
40 "Essay of 1844," p. 221.
41 *LLD*, 1:410–11.
42 See *LLD*, 1:442–3.
43 *Origin*, pp. 445–6. An early report of results is in DAR 205.6:40–4.
44 *Origin*, pp. 445–6.
45 He continued: "Embryos which h. *ve* free life as caterpillars – or external parts may vary" (DAR 205.6:38, dated October 1846). In his copy of Milne Edwards's *Introduction à la Zoologie Générale*, at the mention of parts in which abortive structures are likely (p. 154) he wrote: "Are not these parts last-formed in womb & so exposed to more modifying circumstances" (Darwin's Library).
46 Gould, *Ontogeny and Phylogeny*, pp. 4, 73.
47 See Chapter 5, pp. 126–9.
48 DAR 72:121. Brackets are Darwin's.
49 Karl Ernst von Baer, *Ueber Entwickelungsgeschichte der Thiere*, 1:221–2.
50 Darwin was of course well aware of the importance of embryology in classification. J. V. Thompson's discovery that barnacles are articulates, not molluscs, left little doubt on this score (see Winsor, "Barnacle Larvae"). In the "Sketch of 1842" Darwin gave an evolutionary explanation of the fact (p. 45). But the law that embryos of related forms resemble each other did not prepare Darwin for the claim that the *duration* of embryonic resemblance is a measure of *degree* of affinity.
51 "Essay of 1844," p. 219.
52 DAR 205.6:34, dated November 1844.
53 *Origin*, p. 448.
54 DAR 205.6:27 (undated, mid-1840s) (my emphasis). Darwin at some point canceled the portion I have italicized and the remainder of the sentence in which it occurs.
55 Darwin recognized that evolution will sometimes change the embryo (see "Essay of 1844," p. 228); but he had not anticipated a regular law of embryonic modification.
56 Auguste Brullé, "Recherches sur les Transformations des Appendices dans les Articulés"; DAR 72:123–4 (notes on Brullé, which Darwin read in December 1846).
57 *Natural Selection*, p. 275 (Darwin's emphasis).
58 These are in DAR 205.6.
59 *MLD*, 1:145; *LLD*, 2:7, 39, 133, 162.
60 *Origin*, pp. 457–8. The four sections of Chapter XIII are classification, morphology, embryology, and abortive organs. Paleontology, a part of morphology and of geology, is treated in the chapter on geological succession.
61 E.g., *LLD*, 1:519.
62 Camille Limoges, *La Sélection Naturelle*, pp. 26–85, 142; Janet Browne,

"Darwin's Botanical Arithmetic and the 'Principle of Divergence' "; Frank J. Sulloway, "Geographic Isolation in Darwin's Thinking."

63 Peter Bowler, *Fossils and Progress*. See also Martin J. S. Rudwick, *The Meaning of Fossils*, pp. 218–30. Rudwick, however, places paleontology's contribution to evolution chiefly after 1859 (pp. 245–64).

64 Darwin referred to Owen's "grand views . . . about Palaeotherium & Anoplotherium being prototypes of his two new orders of equal & unequal toed Rum & Pach." (DAR 205.1:47v, dated March 1846). *Cf. Origin*, p. 329. Most of Darwin's paleontological notes are in DAR 205.9; some are in DAR 205.1 and DAR 205.5.

65 Besides the writings of Owen, Carpenter, and Agassiz, which Darwin particularly attended to, the works of two other mid-century paleontologists, H. G. Bronn and F. J. Pictet, contain lists and discussions of many of the most important laws: Pictet, *Traité de Paléontologie*, 1:42–77; Bronn, *Index Palaeontologicus*, 2:809–913.

66 E notebook, pp. 83–4.

67 "Essay of 1844," pp. 229–30. In a note dated November 1854 Darwin wrote: "As in earliest days young no doubt like old, so once there lived an animal like our embryos; but then our embryos no doubt have to a certain extent been modified, it is a chance if such a resemblance c^d be detected" (DAR 205.6:48).

68 Leonard G. Wilson, ed., *Sir Charles Lyell's Scientific Journals on the Species Question*, pp. xliv–xlv; *Origin*, p. 338. See also *MLD*, 1:73–7.

69 *Origin*, p. 330.

70 For instance, in his copy of Hermann Burmeister's *The Organization of Trilobites*, p. 37, Darwin noted, "The earlier geological types present peculiarities of various existing groups passing into one another. Good remark to quote" (Darwin's Library).

71 William B. Carpenter, *Principles of Comparative Physiology*, 4th ed., pp. 107–17. Darwin commented in his copy, "An admirable summary chiefly from Owen on this subject" (Darwin's Library).

72 *Origin*, pp. 330–6, 343. On Darwin's criticism of Huxley, see Dov Ospovat, "Darwin on Huxley and Divergence," p. 216.

73 *LLD*, 2: 39, 57, 131; *MLD*, 1: 140.

74 Peter Bowler's observation is worth repeating, that despite the deficiencies of the fossil record paleontology was never developed into a leading argument against Darwin's theory (*Fossils and Progress*, p. 118).

75 It is important to keep in mind that "descent" and "transmutation" were not necessarily synonymous (see Chapter 5, n. 55). The extent to which Chambers improved his book after the first edition is rarely appreciated. The tenth edition (1853), which Huxley attacked with such venom, was corrected for the author by no less an authority than Carpenter; in it Chambers adopted the main lines of the branching conception from Carpenter's *Comparative Physiology* (see Marilyn Ogilvie, "Robert Chambers and the Successive Revisions of the *Vestiges of the Natural History of Creation*").

76 Charles Darwin, *Autobiography*, p. 124.

77 Alvar Ellegård, *Darwin and the General Reader*, p. 17.

78 H. G. Bronn, "Review of the *Origin of Species*," p. 124. See also Carl Gegenbaur, *Gründzuge der Vergleichenden Anatomie*, 2nd ed. (Leipzig, 1870), p. 19, quoted in William Coleman, "Morphology between Type Concept and Descent Theory," p. 162.

79 See, e.g., *LLD*, 1:519; 2:6–7, 57–8, 155–6, 160–2.

80 This is not only true of Part II. In his discussion of the difficult problem of the gradual evolution of very perfect and complex structures Darwin relied heavily on "transitional states" revealed by recent morphological research: see *Natural Selection*, esp. pp. 350–64.

81 On the rapid acceptance of the theory see David L. Hull, Peter D. Tessner, and Arthur M. Diamond, "Planck's Principle." Although this is not the authors' conclusion, their data indicate that a large majority of the scientists in their sample accepted evolution by 1869 (their study is marred, however, by a peculiar choice of scientists). See also Ellegård, *Darwin and the General Reader*, p. 337.

7. Classification and the principle of divergence

1 Gavin de Beer, ed., "Darwin's Journal," p. 13; *LLD*, 1:362; Charles Darwin, *Autobiography*, p. 118.

2 I should note that a casual remark by Darwin (*LLD*, 2:211) led a number of writers to speculate that he formulated the principle of divergence in 1852 (Gavin de Beer, *Charles Darwin;* p. 140; Camille Limoges, *La Sélection Naturelle*, pp. 134–6; Silvan S. Schweber, "Darwin and the Political Economists"; Frank J. Sulloway, "Geographic Isolation in Darwin's Thinking"; p. 35). But as far as I know the only historian besides me whose determination of the date is based on examination of Darwin's notes on the subject has concluded that the principle of divergence was invented in the summer of 1857, as the culmination of work begun in the fall of 1854 (Janet Browne, "Darwin's Botanical Arithmetic and the 'Principle of Divergence,' 1854–1858"). I will show that what occurred in the summer of 1857 was the discovery of a major new role for a principle Darwin had already developed the year before. But the notes leave no doubt that the period 1854–7 is right, as Janet Browne has said.

3 In Schweber's and Sulloway's articles (see preceding note) it is simply taken for granted that the problem concerned speciation and geographical distribution. See also De Beer, *Charles Darwin*, p. 140; and Vorzimmer, "Darwin's Ecology and Its Influence upon His Theory," pp. 152–3.

4 Darwin, *Autobiography*, p. 120. Limoges, whose main interest was in Darwin's ecological vision of nature, which is expressed chiefly in the section on divergence in the *Origin*, nevertheless stated clearly that classification was the problem that led Darwin to develop the principle, and this without seeing the 1854–8 notes (*Sélection Naturelle*, pp. 131, 134–6).

5 DAR 205.5:151. The quoted note is dated November 1854.

6 C notebook, p. 155.

7 "Essay of 1844," pp. 208–11. That extinction was the sole reason for the existence of groups is an idea that Darwin expressed on many occasions before 1856: "A genus is a real thing but is consequent solely on extinction, if we take all organisms which have lived" (DAR 205.4:40, dated December 1845). Similar notes are scattered through DAR 205.5.

8 "Essay of 1844," p. 212.

9 *LLD*, 2:42.

10 DAR 205.5:120, dated July 1847.

11 The suggestion that species in large genera are more variable than those in small genera became an important part of Darwin's solution in 1854–6.

12 All group-in-group classification is in a sense branching. But from the teleologists' perspective, each group is merely a generalization of the facts of structure (Cuvier's "type"). From the morphologists' point of view, subtypes are a special development of a more general type of structure, and group-in-group classification implies a developmental connection (real or ideal), rather than simply a convenient way of arranging similar forms. See the note Darwin added, probably in 1854, to the "Essay of 1844," p. 198n.

13 DAR 205.5:168, dated December 1856.

14 The reference is probably to Lyell, who in 1856 was still a teleologist and who had been discussing these matters with Darwin since their meeting in April. See Leonard G. Wilson, ed., *Sir Charles Lyell's Scientific Journals on the Species Question*, pp. 3, 52–60, 83–89.

15 DAR 205.5:173–4; "((preexistent))" is Darwin's addition.

16 DAR 205.5:116, headed "Athenaeum 1845 p. 199."

17 DAR 205.1:47, headed "Owen. March 46"; *MLD*, 1:415.

18 On Darwin's prior knowledge of the work, see the references in Rev. Richard Owen, *The Life of Richard Owen*, 1:209, 292.

19 DAR 72:117.

20 *Origin*, p. 111.

21 DAR 205.5:127, dated December 1848.

22 DAR 205.5:149, dated November 1854.

23 DAR 205.5:151, dated November 1854. Ten years earlier Darwin had expressed his agreement with the hypothesis of Swainson, Waterhouse, and Gould that widespread genera usually have species that range widely 'DAR 205.5:97, dated March 31, 1844; see also *LLD*, 1:385–6; *MLD*, 1:102–3).

24 *Origin*, p. 113.

25 *MLD*, 1:84. Sulloway unaccountably inserts the word "botanical" in a passage from Darwin's notes on his statistical work in Schoenherr (insects): "Geographic Isolation in Darwin's Thinking," p. 40.

26 The calculations are in DAR 205.9:291–302.

27 Most of Darwin's speculations from 1844 on concerning the ranges of typical and large, aberrant and small genera are directed toward questions of classification and have their origin in the same assumptions about the causes of variation and about the external conditions that allow

groups to grow large and range widely: "genera which are not typical are only rendered so by the extinction of allied genera, & that implies they are less adapted than other groups of genera to their world & therefore one might expect they would be less widely distributed" (DAR 205.5:97, dated March 31, 1844). Similar statements are scattered throughout the Darwin MSS for this period.

28 DAR 205.9:303–4, dated November 1854. This is published by Robert C. Stauffer in *Natural Selection*, pp. 581–3. Stauffer's transcription contains a number of errors, at least three of which affect the sense of the note (I intend no disparagement of his monumental job of editing *Natural Selection*). Darwin's sheets of calculations show that by "local" he meant genera confined to one of the five great regions into which he divided the world, e.g., "America, whole."

29 See Chapter 4, pp. 103–4, 108–9.

30 *MLD*, 1:82–8 (the correct dates of these two letters to Hooker are November 15, 1854, and December 11, 1854); DAR 205.5:147, dated November 1854. Darwin was anxious to establish his own explanation of fewness of species in aberrant genera as an alternative to a plausible creationist account: "A creationist might say the fact of aberrance shows that they differ from common form, ie form adapted to commonest circumstances, & therefore it is self-evident they w^d not be likely to have many species created on such type. Quite sufficient explanation. But it is necessary for *me* to account for fewness of species, after having shown such to be the case" (DAR 205.9:310).

31 Most of the calculations are in DAR 205.9. Some of the conclusions from them were later added to "Theoretical Geograph. Distrib." and have been published by Stauffer in *Natural Selection*, p. 583.

32 DAR 205.5:147, dated November 1854.

33 "Essay of 1844," pp. 196–7.

34 "The increasing genera are in my notions genera with close species" (DAR 205.9:304v; published by Stauffer in *Natural Selection*, p. 583).

35 B notebook, p. 149.

36 "In genera containing many species, the individual species stand much closer together than in poor genera" (DAR 73:118). See *Natural Selection*, p. 93, OR *LLD*, 1:460. Darwin's account of his reaction to Fries (*Natural Selection*, pp. 93, 145–7) is colored by hindsight; the problem he was working on in 1855 was not to prove that varieties are incipient species, but rather to explain classification.

37 *Natural Selection*, pp. 146–8.

38 Ibid., pp. 93, 145–8.

39 Here for flavor are two extracts from Darwin's sheets of calculations: 1. (based on Alexandre Boreau, *Flore du Centre de la France*, 2 vols. [Paris, 1840]): "of the 413 genera, 100 have one or more species with vars. & these 100 genera include 536 species, so they have on average 5.36. The remaining 313 (with no species having vars. have consequently (1156 − 536) 620 species, or on an average of only 1.98 species to each genus" (DAR 15.2:4). 2. (based on Asa Gray's MS with close species marked):

"The number of genera with 'close species' is 115; according to general scale 1/10 of these ought to have vars. but in fact 33/115=1/3.5 have vars. which shows to demonstration that genera with 'close species' and vars. are connected" (DAR 15.2:19).

40 *MLD*, 1:40; DAR 100:7.
41 *Natural Selection*, p. 231.
42 DAR 205.5:149, dated November 1854.
43 *LLD*, 1:418; *Natural Selection*, pp. 228–33.
44 DAR 205.3:167, dated January 30, 1855; *Natural Selection*, p. 228.
45 DAR 75:8; *Natural Selection*, p. 229; *Origin*, p. 113.
46 DAR 205.3:167, dated January 30, 1855 (the omitted portion is cited above, n. 44).
47 DAR 205.5:171.
48 *Origin*, pp. 116–22: *Natural Selection*, p. 238.
49 All earlier uses of the principle of the division of labor that I have seen occur in discussions of organic progress or degree of development, e.g.: "There is no law of Progression, but time w^d give better chance of sports, & allow more selection; & all the organisms then living an advantage, – a free competition of labour – the result w^d be complicated and more perfect" (DAR 205.9:50, dated November 1854).
50 *Natural Selection*, p. 233; *Origin*, pp. 115–16.
51 This is especially true in the case of Milne Edwards's physiological principle, with which Darwin was well acquainted as early as 1846 (DAR 72:129). When Limoges discussed Milne Edwards and the principle of divergence he had not had access to the sources I am drawing on here (*Sélection Naturelle*, p. 135); nor has Schweber, who discusses the role of Milne Edwards and political-economic thought, made use of them ("Darwin and the Political Economists"). As for the supposed spark, my guess is that on his memorable carriage ride (*Autobiography*, p. 120) he was struck by the sight of the heath and meadow referred to above.
52 The MSS from March 1857 are discussed in the following section.
53 DAR 205.5:157, dated August 19, 1855; DAR 205.5:174, dated May 11, 1856.
54 *Origin*, pp. 411–12.
55 DAR 205.5:174. Cf. the passage quoted in n. 7, above.
56 For the dates of original drafts and additions, see *Natural Selection*, pp. 92–5, 213.
57 DAR 47:86.
58 DAR 205.5:182.
59 DAR 10.2:MS p. "27 & 28" (these first lines of the original p. 28 are canceled). The part I have quoted is preceded by the words "a small & very large scale" and followed by what is published as fol. 27, 28 in *Natural Selection*, p. 250.
60 That Darwin intended to treat divergence more fully later is clear from the following canceled and truncated remark near the conclusion to the original draft of Chapter VI: "I will only add, that owing to the principle of divergence – to which mere allusion has as yet been made, I" (DAR

10.2:MS p. 77 ['58' deleted]). My guess is that p. 58 was the next to the last page of the original draft of Chapter VI. This sentence was canceled and p. 59 discarded when the section on the principle of divergence was added.

61 *Natural Selection*, pp. 174, 219.

62 Ibid., pp. 280–1 (this appears to be part of the original draft).

63 *LLD*, 1:461–2. The difference between the pre- and post-Lubbock methods is discussed in Browne, "Darwin's Botanical Arithmetic," pp. 78–9. The pre-Lubbock method is that used in the calculation based on Boreau given in n. 39, above: The post-Lubbock method is that displayed in *Natural Selection*, pp. 148–55. Karen H. Parshall has shown that even after the Lubbock episode Darwin's procedure was faulty, so that the figures as given in *Natural Selection* do not logically bear on the question Darwin was addressing. Furthermore, she shows that the results given in *Natural Selection* are not statistically significant: "Varieties are Incipient Species: Darwin's Numerical Analysis" (unpublished essay; I am grateful to its author for permission to refer to it).

64 *Natural Selection*, pp. 145–6.

65 Ibid., pp. 243–4.

66 Ibid., p. 165. The same idea is expressed in a letter to Hooker in 1853 (*LLD*, 1:400).

67 *Natural Selection*, p. 249.

68 *LLD*, 1:462–3, 481; *MLD*, 1:99.

69 *MLD*, 1:105–8; DAR 114:227, 228. Darwin discussed Hooker's chief objection in *Natural Selection*, pp. 155–9.

70 *MLD*, 1:109.

71 Darwin, *Autobiography*, p. 120.

72 On the debate between proponents of ecological and geographical speciation, see Ernst Mayr, *Evolution and the Diversity of Life*, pp. 117–217. See also Sulloway, "Geographic Isolation in Darwin's Thinking," pp. 23–65.

73 Darwin wrote Hooker in 1862 that the principle of divergence "was the best point which, according to my notions, I made out, and it has always pleased me" (*MLD*, 1:199).

8. The principle of divergence and the transformation of Darwin's theory

1 DAR 16.2:303, dated July 1846.

2 *Natural Selection*, p. 386.

3 DAR 205.5:151, dated November 1854 (quoted in Chapter 7, p. 176).

4 DAR 205.5:171, dated September 23, 1856 (quoted in Chapter 7, p. 181).

5 DAR 47:86 (quoted in Chapter 7, p. 184).

6 B notebook, p. 20; "Essay of 1884," p. 85.

7 E notebook, p. 135; see also "Essay of 1844," pp. 91, 183–4.

8 DAR 16.2:304, dated June 1844.

9 *Natural Selection*, pp. 260–1, 265–7, 273–4.

10 DAR 114:36, postmarked July 1845; see *MLD*, 1:51.

11 DAR 205.4:19. Darwin observed at the same time, "I must give up my crossing notions & advantage of Paucity of individuals. I must stick to new conditions & especially new groupings of organic beings."

12 DAR 205.9:303 (published in *Natural Selection*, p. 582).

13 This is first clearly stated in the note of September 23, 1856, but Darwin came close to expressing the same idea in a note dated August 19, 1855: "Owing to powers of propagation not only as many individuals crowded together, but 'forms' for more can be supported on same area, when diverse, than when of same species (here discuss cases of many genera on several spots & conditions: Trifolium at Lands end. Larch wood. Coral Isl^d. Hookers facts.) as when many individuals crowded together some will die, so will forms. Creation causes extinctions – like birth of young causes death of old – All classification follows from more distinct forms being supported on same area" (DAR 205.5:157).

14 *Natural Selection*, p. 254; also pp. 257–60, 266.

15 E notebook, p. 115e; "Essay of 1844," pp. 86–7; DAR 47:1, dated June 1840. See also C notebook, p. 25e.

16 DAR 205.3:161, dated November 1854.

17 E notebook, pp. 95–6.

18 Ibid., p. 108.

19 E.g., "Essay of 1844," pp. 90–2. In a note dated August 1849 he stated his theory as follows: "suppose to make species *some change of condition* & isolation were necessary" (my emphasis) (DAR 47:4).

20 "Station indicates the peculiar nature of the locality where each species is accustomed to grow, and has reference to climate, soil, humidity, light, elevation above the sea, and other analogous circumstances" (Charles Lyell, *Principles of Geology*, 5th ed., 2:11 [Bk. III, chap. 5]).

21 This is well exemplified in the "Essay of 1844," pp. 184–7. I have found one note, however, in which it is suggested that changes in a prey-species could cause a predator to vary. (DAR 47:1, dated June 1840). On Darwin's concept of place, see Robert C. Stauffer, "Ecology in the Long Manuscript Version of Darwin's *Origin of Species* and Linnaeus' *Oeconomy of Nature*"; and Camille Limoges, *La Sélection Naturelle*, pp. 131–2. The distinction Peter Vorzimmer makes between station and place does not seem to me to correspond to Darwin's usage either in 1844 or in the *Origin* ("Darwin's Ecology and its Influence upon His Theory," p. 150).

22 *LLD*, 2:8–9.

23 In the "Essay of 1844" he wrote (p. 185): "it is certain that all organisms are nearly as much adapted in their structure to the other inhabitants of their country as they are to its physical conditions." In 1854 he wrote: "No doubt temperature greatest ruling cause of difference in organisms: but the nature of the associated kinds comes at least next" (DAR 205.3:161, dated November 1854).

24 DAR 205.9:315, dated July 27, 1856. He said the same thing to Hooker in November (*LLD*, 1:445). The occasion for both remarks was Darwin's attempt to justify some parts of his theory of the influence of the glacial

period on geographical distribution (see *Natural Selection*, pp. 542–3, 548–9, 557–8).

25 DAR 205.3:167, dated January 30, 1855, part quoted in Chapter 7, pp. 180–1.

26 *Natural Selection*, p. 247.

27 Ibid., pp. 271–2. The expression "organic action and reaction" is also used in the *Origin*, p. 408.

28 *Natural Selection*, pp. 247–8. In the *Origin* Darwin made no dogmatic assertion that the limits have not been reached (on p. 126 he said it was possible that they were reached long ago). But in *Natural Selection* he indicated that he thought the number was still increasing (p. 248), and he repeated the same idea to Lyell (*LLD*, 1:531) and to H. C. Watson: "I fully . . . admit, that the organic relations will tend, with increase of number of species, to go on getting more & more complex, & thus there will be tendency to increase in number of new species . . . We do not know that full number anywhere arrived at" (DAR 47:137v).

29 I suspect it is this that caused Darwin in the second edition of the *Origin* to drop the wedging metaphor, which he had used since first reading Malthus. The wedging metaphor implied strict limits on the number of places and so was inappropriate given Darwin's new conception of place and his belief that their number may increase indefinitely (Morse Peckham, ed., *The Origin of Species by Charles Darwin*, p. 150).

30 "Essay of 1844," pp. 108–9, 183.

31 *Natural Selection*, pp. 271, 273.

32 Limoges, *Sélection Naturelle*, pp. 127–36, 151; and "Darwinisme et Adaptation," pp. 354, 368–9. If we follow Limoges's arguments on the difference between the traditional conception of the economy of nature and Darwin's ecological conception, we must say that the traditional conception persisted as the basis for Darwin's thought until the 1850s. I should in fairness note that Limoges seems really to have recognized this ("Darwinisme et Adaptation," p. 353), but also to have found it difficult to reconcile with his interpretation of the pre-Malthus notebooks.

33 *Natural Selection*, p. 254.

34 *LLD*, 1:397 (June 13, [1849]).

35 Charles Darwin, *A Monograph on the Sub-Class Cirripedia*, 2:185, 188.

36 DAR 205.6:45, dated February 1852 (my emphasis).

37 See Chapter 7, p. 176.

38 Darwin, *Cirripedia*, 2: 184–8. In July 1855 Darwin reported to Hooker that he had begun experiments to induce variability in plants by artificially altering their conditions of life (DAR 114:141).

39 It has recently been suggested that Darwin undertook his work on barnacles in order to prove that variation in nature is common, but there is no evidence for this and a great deal against it (Silvan S. Schweber, "Essay Review: The Young Darwin," p. 186). The available evidence also seems to me to be incompatible with Janet Browne's brief account of the reasons for, timing of, and consequences of the changes in Darwin's views

on variation ("Darwin's Botanical Arithmetic and the 'Principle of Divergence,' pp. 58–9, 62).

40 *Natural Selection*, p. 164. When Darwin began collecting his evidence, and for many years after, his goal was to show the mere "possibility" of selection: "Perhaps I have not been generally cautious enough in selecting cases of variable parts to show possibility of selection." This he observed (in December 1846) in reference to facts he took from Milne Edwards and which he eventually cited in *Natural Selection* (p. 110). There they are part of his argument that there is much variation in nature; but in 1846 he said, "These facts only serve to do away with a priori improbability of variation" (DAR 74:33v).

41 The following passage is indicative of what Darwin expected from his calculations and of how irrelevant the question of changes of conditions had become: "The best territories for my special object, would be those with all the species endemic, for all the species will probably have originated in such areas and where many species of the same genus have been formed, there as a general rule we ought now to find most variation in progress" (*Natural Selection*, p. 158).

42 *Origin*, p. 113.

43 "If we could start with quite similar organisms & bred them for many generations during their whole lives under absolutely similar conditions, the offspring would be absolutely similar" (*Natural Selection*, p. 105).

44 He wrote to Hooker in 1862, "I wish I had . . . started on the fundamental principle of variation being an innate principle" (*MLD*, 1: 199).

45 *Natural Selection*, pp. 218–19. Janet Browne has stated that in *Natural Selection* crossing replaced external change as Darwin's main cause of variation ("Darwin's Botanical Arithmetic," p. 66); but this is not so. The reasons for Darwin's interest in crossing are given in *Natural Selection*, pp. 35, 43.

46 "Essay of 1844," p. 186, n. 1.

47 DAR 45:65, undated.

48 "Some reasons were given in the first chapter for supposing that abundant food might be one main cause of variation under domestication; & I think I shall be able to show hereafter that the species, which are now most vigorous, ranging furthest & abounding most in individuals are those which vary most; & thus we may believe are the best nurtured" (*Natural Selection*, p. 174).

49 See *Natural Selection*, p. 105, where "varied external conditions" has been changed to "varied conditions."

50 Charles Darwin, *The Variation of Animals and Plants under Domestication*, 2: 309–10.

51 D notebook, pp. 167, 175; "Essay of 1844," pp. 84–5. See Chapter 3, pp. 80–1. As late as 1852, after the barnacle work, Darwin still supposed the general rule to be "that parts of little importance alone vary" (DAR 206).

52 *Natural Selection*, pp. 103–33.

53 On this basis one might wish to say (as one of my readers has) that Darwin's theory of variation remained fundamentally the same. But one would still have to recognize radical changes in emphasis. To me it seems that these radical changes are sufficiently important to warrant being called a new theory. The quoted passage is from DAR 45:42. See also *Natural Selection*, pp. 105–6; *Variation under Domestication*, 2:305; *MLD*, 1:197–8.

54 "Mere variability, though the necessary foundation of all modifications, I believe to be almost always present, enough to allow of any amount of selected change" (August 1862) (*LLD*, 2:180). See also *MLD*, 1:171.

55 The note in which this passage occurs is interesting enough to quote at length: "In *large* continent, the individuals of each species from being exposed to variable conditions & from crossing [of] the sub sub-varieties, cannot be so perfectly adapted to conditions, as in island – this must hold good to whole fauna – therefore a species from neighboring islands, from being more closely conformable to these conditions [Darwin placed a question mark in the margin at this point], will exceed & beat out the continental species. –

"In old times there have been genera, families, &c. &c.

"If every species had heirs, the number would be infinitely great in times to come. – but if we suppose the number of individuals & diversions equal in two times, only a few individuals can have descendants, hence much extinction in proportion to metamorphosis: hence gradation not or only seldom discoverable in thick stratum. N.B. The adaptation of any species in same country, so that the perfection of adaptation in two countries may possibly vary. – : under a changing climate, species less conformable, than in old fixed climate" (DAR 205.9:81, undated [the paper suggests this was written ca. 1840]).

56 Except, of course, in the sense that external changes may make a group less well adapted (or better adapted) than some other group, in which case it will be on its way to extinction (or expansion). This idea was present in Darwin's earliest speculations: B notebook, pp. 37–8, 205–6; C notebook, p. 151; "Essay of 1844," p. 209.

57 DAR 205.9:304 (published in *Natural Selection*, p. 582 [there are no question marks in the original]).

58 *Origin*, p. 201.

59 Ibid., p. 102.

60 *Natural Selection*, p. 219. Cf. "Essay of 1844," p. 153; and see Chapter 3, pp. 74–5.

61 The tone of his letter to Hooker on the subject (September 11, 1857) suggests that relative adaptation was still a new and exciting concept for Darwin: "I have just been writing an audacious little discussion to show that organic beings are not perfect, only perfect enough to struggle with their competitors" (DAR 114:211).

62 *Natural Selection*, pp. 247–8.

63 Ibid., p. 225; Darwin, *Origin*, p. 84. On Darwin's subsequent use of the

language of perfect adaptation, see *MLD*, 1:202 (regarding his book on orchids).

64 Susan F. Cannon, *Science in Culture*, p. 248; Walter F. Cannon, "History in Depth," p. 32. The theological ideas Cannon has in mind are not precisely the same ones I have been discussing (see Chapter 3, n. 66), nor do we agree on the meaning of perfect adaptation in the period: see Walter F. Cannon, "The Bases of Darwin's Achievement," p. 128; W. Faye Cannon, "Charles Lyell, Radical Actualism, and Theory," pp. 118, 120 n. 28. For a much broader discussion, along the same lines as Cannon's, of the importance of theological traditions in understanding Darwin's thought and that of his contemporaries, see James R. Moore, *The Post-Darwinian Controversies*, esp. pp. 217–345.

65 Part of the fascination of *Natural Selection* is that it reveals Darwin's thought in a state of transition; on one page he speaks of relative adaptation, on another, perfect adaptation; food has become an important cause of variation, but not yet the most important, etc.

9. Natural selection and "natural improvement"

1 DAR 205.9:153 (Darwin read *Old Red Sandstone* in June 1842).
2 See Chapter 6, p. 147.
3 While many have recognized that Darwin believed progress was an inevitable consequence of natural selection, others have denied that he had any theory of progress. In the first category are John C. Greene, "Darwin as a Social Evolutionist"; Peter J. Bowler, "The Changing Meaning of 'Evolution,'" p. 109; Maurice Mandelbaum, *History, Man, & Reason*, p. 82. In the second category are M. J. S. Hodge, "England," p. 16; Stephen J. Gould, *Ever since Darwin*, pp. 36–7.
4 Morse Peckham, ed., *The Origin of Species by Charles Darwin*, p. 221 (3rd ed., 1861). There is no satisfactory treatment of Darwin's general theory of progress. On his views on human progress see Greene, "Darwin as a Social Evolutionist."
5 A note from about the mid-1850s is of interest in this regard: "If physical mutations of world go in cycle (perhaps endless combinations) yet organic not so, for each form bears impress of ancestors & is related not only to the cyclical external conditions, but to the progressive organic world" (DAR 205.5:164v).
6 B notebook, pp. 18–19.
7 Ibid., p. 78; D notebook, pp. 49, 57, 170.
8 "Essay of 1844," p. 230. See also *Origin*, p. 338.
9 DAR 48 (ser. 1):82, undated (probably 1856–8).
10 B notebook, pp. 108e–9, 204–7; E notebook, p. 60.
11 On Lamarck see Richard W. Burkhardt, Jr., *The Spirit of System*, p. 146.
12 B notebook, pp. 49, 169, 214–15; C notebook, p. 74. In C notebook, pp. 78–9, Darwin suggested that as long as circumstances were favorable for

man, man might be produced again from monkeys, but this was not his usual conclusion.

13 See E notebook, pp. 68–9. The inevitability of something like man seems also to be suggested in Charles Darwin, *Autobiography*, p. 92.

14 D notebook, pp. 49, 58; E notebook, pp. 70, 95–7; DAR 47:64–5; "Essay of 1844," p. 227.

15 See William Buckland, *Geology and Mineralogy considered with reference to Natural Theology*, 1:107–8n.

16 See Chapter 5, p. 120.

17 Richard Owen, *Lectures on the Comparative Anatomy and Physiology of the Invertebrate Animals*, p. 365.

18 Henri Milne Edwards, *Introduction à la Zoologie Générale*, pp. 35–57.

19 Charles Lyell, "Anniversary Address of the President"; [Richard Owen], "Lyell – on Life and its Successive Development."

20 See Chapter 5, n. 52. Again, many of these developmental interpretations of organic succession made no attempt to provide a causal theory of succession.

21 Owen Collection, Correspondence, vol. 22, p. 268 (letter, Sedgwick to Owen, 1860), British Museum (Natural History); Roderick I. Murchison, *Siluria*, pp. 466–7.

22 [Robert Chambers], *Vestiges of the Natural History of Creation*, 10th ed., p. 156.

23 On Spencer see Chapter 5, n. 4; see also Herbert Spencer, *An Autobiography*, 1:445; and Stephen J. Gould, *Ontogeny and Phylogeny*, pp. 112–14.

24 DAR 205.9:200, dated March 1845.

25 When Milne Edwards proposed comparing the degree of development of a single animal at different ages and of one species with neighboring species, Darwin commented, "Best way of putting superiority – though each perfectly (?) (can young be said to be perfectly ?) adapted to conditions" (Darwin's Library, Milne Edwards, *Introduction à la Zoologie Générale*, p. 21).

26 *MLD*, 1:413–16.

27 "Why few legs sh^d serve function better than many I do not see – Probably the complication of one or a few pairs, by compensation, causes very few to be perfected or created. Hence fewness is a consequence of perfection" (DAR 72:129v). "Observations sur la Circulation" appeared in *Annales des Science Naturelles*, 3rd ser., 3 (1845): 257–88.

28 DAR 72:118–19.

29 *MLD*, 1:76–77. On Darwin's reading of von Baer, see Charles Darwin, *Monograph on the Sub-Class Cirripedia*, 2:19–20, where definitions of high and low are discussed. Darwin and Hooker also discussed this topic in 1852 (DAR 205.9:246).

30 One such speculation is quoted in Chapter 7, n. 49. On Darwin's reservations, see Darwin, *Cirripedia*, 2:19–20.

31 *LLD*, 1:498 (letter to Hooker, December 1858).

32 *MLD*, 1:114 (letter to Hooker, December 1858). When Lyell, committed to the traditional conception of adaptation, objected, Darwin replied: "I

have altered the sentence about the Eocene fauna being beaten by the recent, thanks to your remark. [He added the phrase "under a nearly similar climate"; *Origin*, p. 337.] But I imagined it would have been clear that I supposed the climate to be nearly similar . . . Not that I think climate nearly so important as most naturalists seem to think. In my opinion no error is more mischievous than this"(*LLD*, 1:523).

33 In his notes on Milne Edwards's "Observations sur la Circulation," Darwin remarked that "one can see how the physiological division of labour profits in such cases, for digestion can hardly go on so vigorously when mixed with water for äeration: äeration imperfect & the circulation cannot be vigorous without disturbing digestion" (DAR 72:129).

34 *Natural Selection*, pp. 234–5.

35 *MLD*, 1:115; DAR 205.3:194, undated, headed "this will come under Geograph Distrib."

36 *Natural Selection*, p. 355. See also Darwin's remarks to Lyell regarding the chapter on organic progress in the tenth edition of Lyell's *Principles:* "You do not allude to one very striking point enough, or at all – viz., the classes having been formerly less differentiated than they now are; and this specialisation of classes must, we may conclude, fit them for different general habits of life as well as the specialisation of particular organs" (*MLD*, 1:272).

37 DAR 48:82. See Greene, "Darwin as a Social Evolutionist," p. 5.

38 There he insisted only that recent forms must be competitively higher than ancient forms. His belief in progress is suggested primarily in his discussions of embryology and in the conclusion to the book (*Origin*, pp. 336–8, 449, 490).

39 On Darwin and the reception of *Vestiges* see Frank N. Egerton, "Refutation and Conjecture"; on Huxley's nonprogressionism see Michael Bartholomew, "Huxley's Defence of Darwin."

40 See Dov Ospovat, "Lyell's Theory of Climate," p. 336.

41 *LLD*, 1:528, 530–1; 2:6–7.

42 In the third edition he added further: "and consequently, in most cases, to what must be regarded as an advance in organisation" (Peckham, ed., *Origin of Species by Darwin*, p. 271).

43 Ibid., p. 547.

44 Ibid., p. 222.

45 Ibid., pp. 222–5, 547–51. Even in these remarks, however, Darwin indicated that the presently existing lower forms are probably higher than those that formerly occupied similar places in the economy of nature (ibid., p. 223: *MLD*, 1:142).

46 *MLD*, 1:164; Peckham, ed., *Origin of Species by Darwin*, pp. 224–5; ibid., pp. 548–9, records Darwin's increasing confidence in successive editions.

47 *Origin*, p. 490.

48 Mandelbaum, *History, Man, & Reason*, pp. 85–7. Cf. the more subtle argument in James R. Moore, *The Post-Darwinian Controversies*, pp. 157–61.

49 It should be recalled, however, that Darwin's nature was not (in his view)

quite so unpleasant as it is often portrayed. In his *Autobiography* he said, "According to my judgment happiness decidedly prevails [over suffering in nature], though this would be very difficult to prove. If the truth of this conclusion be granted, it harmonises well with the effects which we might expect from natural selection" (p. 88).

50 Greene, "Darwin as a Social Evolutionist," p. 26. Darwin's hopes for the future are well indicated by his remark to Lyell that it would be "an infinite satisfaction" to him to believe that "mankind will progress to such a pitch that we should [look] back at [ourselves] as mere Barbarians" (*MLD*, 2:30).

51 Assuming, that is, that the process of transmutation and therefore increasing perfection would begin at all. Darwin expressed perplexity as to why the simplest forms should ever have changed (Peckham, *Origin of Species by Darwin*, p. 225; *MLD*, 1:164).

52 DAR 48:82; Peckham, *Origin of Species by Darwin*, p. 222. See also *LLD*, 1:528.

53 Charles Darwin, *The Descent of Man and Selection in Relation to Sex*, pp. 99, 130 (beginning of chap. 4, beginning of chap. 5).

54 See, e.g., J. D. Y. Peel, *Herbert Spencer*, p. 142.

55 An omniscient creator could of course know beforehand every detail of his evolving universe. But if so, why work through the wasteful and irrational process of chance variations and selection? If not, can there be said to be an omniscient creator? See Darwin, *The Variation of Animals and Plants under Domestication*, 2:514–16; and see the discussion in Moore, *Post-Darwinian Controversies*, pp. 273–7, 290–2.

56 Darwin, *Autobiography*, p. 93; *MLD*, 1:321.

57 B notebook, p. 216; see also p. 49.

58 Darwin, *Autobiography*, p. 92.

59 Hodge, "England," p. 16; see also Gould, *Ever since Darwin*, p. 37.

60 Quoted in John C. Greene, "Reflections on the Progress of Darwin Studies," p. 256n.

61 Ibid. See also Bowler, "Changing Meaning of 'Evolution,'" pp. 108–9. In Lamarck's theory the necessity of adaptation to various environments could deflect organisms from their upward course.

62 Gould, *Ever since Darwin*, pp. 36–7.

63 For example, "Begin with what is meant by gradations." DAR 205.9.

64 Darwin, *Cirripedia*, 2:19–20.

65 *LLD*, 1:498; *MLD*, 1:114–15.

66 [J. B. Mozley], "The Argument of Design," *Quarterly Review*, 127 (1869);172: "Supposing Mr. Darwin's theory of Progress to be true" (quoted by Robert Young in "Darwin's Metaphor," p. 484).

67 Instead he wrote, "I can say nothing against your side, but I have an 'inner consciousness' (a highly philosophical style of arguing!) that something could be said against you; for I cannot help hoping that you are not quite as right as you seem to be. Finally, I cannot tell why, but when I finished your Address I felt convinced that many would infer that you were dead

against change of species, but I clearly saw that you were not. I am not very well, so good-night, and excuse this horrid letter" (*MLD*, 2:234).

Conclusion: The development of Darwin's theory as a social process

1 Walter F. Cannon, "The Bases of Darwin's Achievement," pp. 109–34; W. Faye Cannon, "The Whewell-Darwin Controversy," pp. 379–80; Camille Limoges, "Darwinisme et Adaptation," pp. 357–8, 363–5; Limoges, *La Sélection Naturelle*, pp. 31–47.
2 John Brooke has suggested other functions of natural theology, relating especially to the professional and social needs of scientists: "The Natural Theology of the Geologists."
3 See Chapter 1, n. 90.
4 "A scientific worker," Joseph Needham remarked, "is necessarily the child of his time and *the inheritor of the thought of many generations*" (my emphasis) ("Limiting Factors in the Advancement of Science as Observed in the History of Embryology," p. 1).
5 I do not mean to say that these problems are not worthy of attention, but merely that they are difficult of solution. Nor is it certain that we should normally expect to find any very close correlations. See the comments by Roy Porter, "The Industrial Revolution and the Rise of the Science of Geology," p. 342.
6 See Walter E. Houghton, *The Victorian Frame of Mind, 1830–1870*, pp. 27–9, where it is argued that there was a revival of faith in progress in the 1830s.
7 See Robert Young, "The Historiographic and Ideological Contexts of the Nineteenth-Century Debate on Man's Place in Nature," pp. 384, 428; and R. C. Lewontin, "Adattamento."
8 Because in its emphasis on ecological isolation it obscured the now generally admitted importance of geographical isolation. See Chapter 7, n. 72.

Bibliography

Manuscript collections

Darwin Archive, Cambridge University Library.
Thomas Henry Huxley Papers, Archives, Imperial College of Science and Technology, London.
Richard Owen Collection, British Museum (Natural History).
Richard Owen Manuscripts, Royal College of Surgeons, London.

Published sources

Agassiz, Louis. *Monographie des Poissons Fossiles du Vieux Grès Rouge ou Système Dévonien (Old Red Sandstone)*. Neuchâtel, 1844.
    "On the Differences between Progressive, Embryonic, and Prophetic Types." *Proceedings of the American Association for the Advancement of Science, 1849*, pp. 432–8.
An Essay on Classification. London: Longman, Brown, Green, Longmans, and Roberts and Trübner and Co., 1859.
Albury, W. R., and D. R. Oldroyd. "From Renaissance Mineral Studies to Historical Geology, in the Light of Michel Foucault's *The Order of Things*." *British Journal for the History of Science*, 10 (1977); 187–215.
Allen, David E. *The Naturalist in Britain: A Social History*. Harmondsworth, Middlesex: Penguin Books, 1978 (first published, 1976).
Appel, Toby A. "The Cuvier-Geoffroy Debate and the Structure of Nineteenth Century French Zoology." Ph.D. dissertation, Princeton University, 1975.
    "Henri de Blainville and the Animal Series: A Nineteenth Century Chain of Being." *Journal of the History of Biology*, 13 (1980): 291–319.
Audouin, Victor. "Recherches Anatomiques sur le Thorax des Animaux Articulés et celui des Insectes Hexapodes en Particulier." *Annales des Sciences Naturelles*, 1 (1824): 97–135, 416–32.
Baden-Powell, R. S. S. *Essays on the Spirit of the Inductive Philosophy, the Unity of Worlds, and the Philosophy of Creation*. London: Longman, Brown, Green, and Longmans, 1855.
Baer, Karl Ernst von. "Ueber das äussere und innere Skelet." J. F. Meckel's *Archiv für Anatomie und Physiologie*, 1 (1826): 327–76.
    "Beiträge zur Kenntniss der Niedern Thiere." *Nova Acta Physioco-Medica*

Bibliography

Academiae Caesarae Leopoldino-Carolinae Naturae Curiosorum, 13, part 2 (1827): 523–762.

Ueber Entwickelungsgeschichte der Thiere. 2 vols. Königsberg: Borntrager, 1828–37.

"Das Allgemeinste Gesetz der Natur in aller Entwickelung." In von Baer, *Reden Gehalten in Wissenschaftlichen Versammlungen.* Vol. 1. St. Petersburg: Karl Röttger, 1864.

Barlow, Nora. "Darwin's Ornithological Notes." *Bulletin of the British Museum (Natural History),* Historical Series, 2 (1963): 201–78.

Barlow, Nora, ed. *Charles Darwin's Diary of the Voyage of H.M.S. "Beagle."* Cambridge: Cambridge University Press, 1933.

Barnes, Barry. *Interests and the Growth of Knowledge.* London: Routledge & Kegan Paul, 1977.

Barrett, Paul H., ed. "Darwin's Early and Unpublished Notebooks" [the M and N notebooks]. In Howard E. Gruber, *Darwin on Man.* New York: E. P. Dutton, 1974.

The Collected Papers of Charles Darwin. 2 vols. Chicago: University of Chicago Press, 1977.

Barry, Martin. "On the Unity of Structure in the Animal Kingdom." *Edinburgh New Philosophical Journal,* 22 (1836–7): 116–41, 345–64.

Bartholomew, Michael. "Lyell and Evolution: An Account of Lyell's Response to the Prospect of an Evolutionary Ancestry for Man." *British Journal for the History of Science,* 6 (1973): 261–303.

"Huxley's Defence of Darwin." *Annals of Science,* 32 (1975): 525–35.

Bell, Charles. *The Hand: Its Mechanism and Endowments as Evincing Design.* London: William Pickering, 1834.

Blumenbach, Johann F. *An Essay on Generation,* trans. by A. Crichton. London: T. Cadell ct al., 1793.

Blyth, Edward. "Observations on the Various Seasonal and Other External Changes which Regularly Take Place in Birds . . . ; with Remarks . . . upon the Natural System of Arrangement." *Magazine of Natural History,* 9 (1836): 393–409, 505–14.

Bowler, Peter J. "Darwin's Concepts of Variation." *Journal of the History of Medicine and Allied Sciences,* 29 (1974): 196–212.

"The Changing Meaning of 'Evolution.' " *Journal of the History of Ideas,* 36 (1975): 95–114.

Fossils and Progress. New York: Science History Publications, 1976.

"Malthus, Darwin, and the Concept of Struggle." *Journal of the History of Ideas,* 37 (1976): 631–50.

Boyle, Robert. *The Works of the Honourable Robert Boyle.* New ed. 6 vols. London: W. Johnston et al., 1772.

Brandon, Robert N. "Adaptation and Evolutionary Theory." *Studies in the History and Philosophy of Science,* 9 (1978): 181–206.

Bronn, H. G. *Index Palaeontologicus.* Vol. 2. Stuttgart: G. Schweizerbart, 1849.

"Review of the *Origin of Species,*" trans. David L. Hull. In David L. Hull, *Darwin and His Critics.* Cambridge, Mass.: Harvard University Press, 1973.

Bibliography

Brooke, John H. "Natural Theology and the Plurality of Worlds: Observations on the Brewster–Whewell Debate." *Annals of Science,* 34 (1977): 221–86.

"Richard Owen, William Whewell, and the *Vestiges.*" *British Journal for the History of Science,* 10 (1977): 132–45.

"The Natural Theology of the Geologists: Some Theological Strata." In *Images of the Earth: Essays in the History of the Environmental Sciences,* ed. L. J. Jordanova and Roy S. Porter. Chalfont St. Giles: British Society for the History of Science, 1979.

Browne, Janet. "The Charles Darwin–Joseph Hooker Correspondence: An Analysis of Manuscript Resources and Their Use in Biography." *Journal of the Society for the Bibliography of Natural History,* 8, part 4 (1978): 351–66.

"Darwin's Botanical Arithmetic and the 'Principle of Divergence,' 1854–1858." *Journal of the History of Biology,* 13 (1980): 53–89.

Brullé, Auguste. "Recherches sur les Transformations des Appendices dans les Articulés." *Annales des Sciences Naturelles,* 3rd ser., 2 (1844): 271–374.

Buckland, William. "Lecture." *Report of the British Association for the Advancement of Science, 1832,* pp. 104–10.

Geology and Mineralogy considered with reference to Natural Theology. 2 vols. London: William Pickering, 1836. 2nd ed., 1837.

Burkhardt, Richard W., Jr. "The Inspiration of Lamarck's Belief in Evolution." *Journal of the History of Biology.* 5 (1972): 413–38.

The Spirit of System: Lamarck and Evolutionary Biology. Cambridge, Mass.: Harvard University Press, 1977.

"Closing the Door on Lord Morton's Mare: The Rise and Fall of Telegony." *Studies in History of Biology,* 3 (1979): 1–21.

Burmeister, Hermann. *The Organization of Trilobites,* trans. Thomas Bell and Edward Forbes. London: Ray Society, 1846.

Cain, A. J. "The Perfection of Animals." In *Viewpoints in Biology,* ed. J. D. Carthy and C. L. Duddington, 3 (1964): 36–63.

Cannon, Susan F. *Science in Culture: the Early Victorian Period.* New York: Dawson and Science History Publications, 1978.

Cannon, W. Faye. "Charles Lyell, Radical Actualism, and Theory." *British Journal for the History of Science,* 9 (1976): 104–20.

"The Whewell–Darwin Controversy." *Journal of the Geological Society,* 132 (1976): 377–84.

Cannon, Walter F. "The Problem of Miracles in the 1830's." *Victorian Studies.* 4 (1960): 5–32.

"The Bases of Darwin's Achievement: A Revaluation." *Victorian Studies,* 5 (1961): 109–34.

"History in Depth: The Early Victorian Period." *History of Science,* 3 (1964): 20–38.

Carpenter, William B. *Principles of General and Comparative Physiology.* London: John Churchill, 1839. 2nd ed., 1841.

Principles of Physiology, General and Comparative. 3rd ed. Rev. title. London: John Churchill, 1851.

Principles of Comparative Physiology. 4th ed. Rev. title. London: John Churchill, 1854.

[Carpenter, William B.]. "Physiology an Inductive Science." Review of *A History of the Inductive Sciences,* by William Whewell. *British and Foreign Medical Review,* 5 (1838): 317–42.

Cassirer, H. W. *A Commentary on Kant's Critique of Judgment.* New York: Barnes and Noble, 1970 (first published, 1938).

[Chambers, Robert]. *Vestiges of the Natural History of Creation.* 10th ed. London: John Churchill, 1853.

Vestiges of the Natural History of Creation. Reprint of the 1st ed. (1844), with an introduction by Gavin de Beer. New York: Humanities Press, 1969.

Coleman, William. *Georges Cuvier, Zoologist.* Cambridge, Mass.: Harvard University Press, 1964.

"Limits of the Recapitulation Theory: Carl Friedrich Kielmeyer's Critique of the Presumed Parallelism of Earth History, Ontogeny, and the Present Order of Organisms." *Isis,* 64 (1973): 341–50.

"Morphology between Type Concept and Descent Theory." *Journal of the History of Medicine and Allied Sciences,* 31 (1976): 149–75.

Corsi, Pietro. "The Importance of French Transformist Ideas for the Second Volume of Lyell's *Principles of Geology.*" *British Journal for the History of Science,* 11 (1978): 221–44.

Cuvier, Georges. *Le Règne Animal.* Vol. 1. Paris: Deterville, 1817.

Recherches sur les Ossemens Fossiles. New ed. Vol. 1. Paris: Dufour et D'Ocagne, 1821.

Cuvier, Georges, and A. Valenciennes. *Histoire Naturelles des Poissons.* Vol. 1. Paris: Levrault, 1828.

Darwin, Charles. *Journal of Researches into the Geology and Natural History of the Various Countries Visited by H.M.S. Beagle.* London: Henry Colburn, 1839 [facs. reprint, New York: Hafner, 1952].

A Monograph on the Fossil Lepadidae. In vol. 5 of Paleontographical Society, *Monographs,* 1851.

A Monograph on the Sub-Class Cirripedia. 2 vols. London: Ray Society, 1851–4.

On the Origin of Species. London: John Murray, 1859.

On the Various Contrivances by which British and Foreign Orchids are Fertilised by Insects, and on the Good Effects of Intercrossing. London: John Murray, 1862.

The Variation of Animals and Plants under Domestication. 2 vols. New York: Orange Judd & Company, 1868.

The Descent of Man and Selection in Relation to Sex. 2nd ed. New York: D. Appleton, 1899.

The Voyage of the Beagle. Garden City, N.Y.: Doubleday, 1962 (the 1860 text of the 2nd [1845] ed.).

The Autobiography of Charles Darwin, ed. Nora Barlow. New York: W. W. Norton, 1969 (1958).

Charles Darwin's Natural Selection: Being the Second Part of His Big Species Book Written from 1856 to 1858, ed. Robert C. Stauffer. Cambridge: Cambridge University Press, 1975.

Darwin, Francis, ed. *The Life and Letters of Charles Darwin.* 2 vols. New York: D. Appleton, 1888.

The Foundations of the Origin of Species: Two Essays written in 1842 and 1844, by Charles Darwin. Cambridge: Cambridge University Press, 1909.

Darwin, Francis, and A. C. Seward, eds. *More Letters of Charles Darwin.* 2 vols. New York: D. Appleton, 1903.

De Beer, Gavin, ed. "Darwin's Journal." *Bulletin of the British Museum (Natural History),* Historical Series, 2 (1959): 1–21.

Charles Darwin: Evolution by Natural Selection. Garden City, N.Y.: Doubleday, 1967.

De Beer, Gavin, M. J. Rowlands, and B. M. Skramovsky, eds. "Darwin's Notebooks on Transmutation of Species." 6 parts. *Bulletin of the British Museum (Natural History),* Historical Series, 2 (1960–1): 23–200; 3 (1967): 129–76.

De Candolle, Alphonse. *Géographie Botanique Raisonée.* 2 vols. Paris: Victor Masson, 1855.

De Candolle, Augustin Pyramus. *Essai Elémentaire de Géographie Botanique.* Extract from vol. 18 of *Dictionnaire des Sciences Naturelles,* 1820.

De la Beche, Henry T. *Researches in Theoretical Geology.* London: Charles Knight, 1834.

Egerton, Frank N. "Studies of Animal Populations from Lamarck to Darwin." *Journal of the History of Biology,* 1 (1968): 225–59.

"Humboldt, Darwin, and Population." *Journal of the History of Biology,* 3 (1970): 325–60.

"Refutation and Conjecture: Darwin's Response to Sedgwick's Attack on Chambers." *Studies in the History and Philosophy of Science,* 1 (1970): 176–83.

"Darwin's Method or Methods?" Review of M. T. Ghiselin, *The Triumph of the Darwinian Method. Studies in the History and Philosophy of Science,* 2 (1971): 281–6.

"Changing Concepts of the Balance of Nature." *Quarterly Review of Biology,* 48 (1973): 322–50.

Ellegård, Alvar. *Darwin and the General Reader.* Göteborg, 1958 (*Göteborgs Universitets Arsskrift,* vol. 64).

Foucault, Michel. *The Order of Things.* New York: Random House, Pantheon Books, 1970.

Freeman, Derek. "The Evolutionary Theories of Charles Darwin and Herbert Spencer." *Current Anthropology,* 15 (1974): 211–37.

Freeman, R. B. *The Works of Charles Darwin: An Annotated Bibliographical Handlist.* Folkestone, Kent: Dawson and Archon Books, The Shoestring Press, 1977.

Charles Darwin: A Companion. Folkestone, Kent: Dawson and Archon Books, The Shoestring Press, 1978.

Freeman, R. B., and P. J. Gautrey. "Darwin's *Questions about the Breeding of Animals,* with a Note on *Queries about Expression." Journal of the Society for the Bibliography of Natural History,* 5, part 3 (October 1969): 220–5.

Bibliography

Gale, Barry G. "Darwin and the Concept of a Struggle for Existence: A Study in the Extrascientific Origins of Scientific Ideas." *Isis,* 63 (1972): 321–44.

Garfinkle, Norton, "Science and Religion in England, 1790–1800." *Journal of the History of Ideas,* 16 (1955): 376–88.

Geddes, Patrick. "Morphology." *Encyclopaedia Britannica,* 9th ed.

Geison, Gerald L. *Michael Foster and the Cambridge School of Physiology: The Scientific Enterprise in Late Victorian Society.* Princeton, N.J.: Princeton University Press, 1978.

Geoffroy Saint-Hilaire, Etienne. *Philosophie Anatomique.* 2 vols. Paris, 1818–22.
Principes de Philosophie Zoologique. Paris: Pichon et Didier, 1830.

Ghiselin, Michael T., and Linda Jaffe. "Phylogenetic Classification in Darwin's *Monograph on the Sub-Class Cirripedia.*" *Systematic Zoology,* 22 (1973): 132–40.

Gillespie, Neal C. *Charles Darwin and the Problem of Creation.* Chicago: University of Chicago Press, 1979.

Gillispie, Charles C. *Genesis and Geology.* New York: Harper & Row, 1959 (first published, 1951).

Gould, Stephen J. *Ever since Darwin.* New York: W. W. Norton, 1977.
Ontogeny and Phylogeny. Cambridge, Mass.: Harvard University Press, 1977.

Gray, Asa. *Darwiniana,* ed. A. Hunter Dupree. Cambridge, Mass.: Harvard University Press, 1963.

Gray, J. E. "Arrangement of Reptiles." *Proceedings of the Zoological Society of London,* Part V, 1837, pp. 131–2.

Greene, John C. "The Kuhnian Paradigm and the Darwinian Revolution in Natural History." In *Perspectives in the History of Science and Technology,* ed. Duane H. D. Roller, Norman: University of Oklahoma Press, 1971.
"Reflections on the Progress of Darwin Studies." *Journal of the History of Biology,* 8 (1975): 243–73.
"Darwin as a Social Evolutionist." *Journal of the History of Biology,* 10 (1977): 1–27.

Gruber, Howard E. *Darwin on Man: A Psychological Study of Scientific Creativity.* New York: E. P. Dutton, 1974.

Heimann, P. M. "Science and the English Enlightenment." *History of Science,* 16 (1978): 143–51.

Herbert, Sandra. "Darwin, Malthus, and Selection." *Journal of the History of Biology,* 4 (1971): 209–17.
"The Place of Man in the Development of Darwin's Theory of Transmutation." 2 parts. *Journal of the History of Biology,* 7 (1974): 217–58; 10 (1977): 155–227.

Himmelfarb, Gertrude. *Darwin and the Darwinian Revolution.* New York: W. W. Norton, 1968 (1959).

Hodge, M. J. S. "England." In *The Comparative Reception of Darwinism,* ed. Thomas F. Glick. Austin: University of Texas Press, 1972.
"The Structure and Strategy of Darwin's 'Long Argument.'" *British Journal for the History of Science,* 10 (1977): 237–46.
"Origins and Species I: Darwin on Extinction and the Laws of Life, 1835–7." Unpublished typescript [1980].

Hofsten, Nils von. "Linnaeus's Conception of Nature." *Kungl. Vetenskaps-Societetens Arsbok 1957.* Uppsala, 1958.

Hooykaas, R. "The Parallel between the History of the Earth and the History of the Animal World." *Archives Internationales d'Histoire des Sciences,* 10 (1957): 3–18.

Catastrophism in Geology. Amsterdam: North Holland Publishing Co., 1970.

Houghton, Walter E. *The Victorian Frame of Mind, 1830–1870.* New Haven, Conn.: Yale University Press, 1957.

Hull, David L. "The Metaphysics of Evolution." *British Journal for the History of Science,* 3 (1967): 309–37.

Hull, David L., Peter D. Tessner, and Arthur M. Diamond. "Planck's Principle." *Science,* 202 (November 17, 1978): 717–23.

Huxley, Leonard. *Life and Letters of Sir Joseph Dalton Hooker.* 2 vols. London: John Murray, 1918.

Huxley, Thomas H. *American Addresses.* London: Macmillan, 1877.

"The coming of Age of 'The Origin of Species'" (1880). In Huxley, *Darwiniana: Essays.* New York: D. Appleton, 1897.

The Scientific Memoirs of Thomas Henry Huxley, ed. Michael Foster and E. Ray Lankester. 4 vols. and supp. London: Macmillan, 1898–1903.

Huxley, Thomas H., trans. "Fragments relating to Philosophical Zoology. Selected from the Works of K. E. von Baer." In *Scientific Memoirs,* ed. Arthur Henfrey and T. H. Huxley. London: Taylor and Francis, 1853.

Jacob, J. R. "The Ideological Origins of Robert Boyle's Natural Philosophy." *Journal of European Studies,* 2 (1972): 1–21.

Jacob, M. C. *The Newtonians and the English Revolution, 1689–1720.* Ithaca, N.Y.: Cornell University Press, 1976.

Kant, Immanuel. *Kritik der Urtheilskraft.* In vol. 5 of *Gesammelte Schriften.* Berlin: Georg Reimer, 1913.

Kirby, William. *On the Power, Wisdom, and Goodness of God, as Manifested in the Creation of Animals, and in their History, Habits, and Instincts.* New ed. 2 vols. London: Henry G. Bohn, 1853.

Kohn, David. "Theories to Work By: Rejected Theories, Reproduction, and Darwin's Path to Natural Selection." *Studies in History of Biology,* 4 (1980): 67–170.

Kottler, Malcolm J. "Charles Darwin's Biological Species Concept and Theory of Geographic Speciation: The Transmutation Notebooks." *Annals of Science,* 35 (1978): 275–97.

Lamarck, J. B. *Histoire Naturelle des Animaux sans Vertèbres.* Vol. 1. Paris, 1815.

Larson, James L. *Reason and Experience: The Representation of Natural Order in the Work of Carl von Linné.* Berkeley: University of California Press, 1971.

"Vital Forces: Regulative Principles or Constitutive Agents? A Strategy in German Physiology, 1786–1802." *Isis,* 70 (1979):235–49.

Lawrence, Philip J. "Charles Lyell versus the Theory of Central Heat: A Reappraisal of Lyell's Place in the History of Geology." *Journal of the History of Biology,* 11 (1978): 101–28.

Le Mahieu, Daniel L. "Malthus and the Theology of Scarcity." *Journal of the History of Ideas,* 40 (1979): 467–74.

Bibliography

Lenoir, Timothy. "Generational Factors in the Origin of *Romantische Naturphi-losophie*." *Journal of the History of Biology*, 11 (1978): 57–100.

"The Göttingen School in the Development of Transcendental Naturphilo-sophie in the Romantic Era." *Studies in History of Biology*, 5 (1981): 110–215.

Levin, Samuel M. "Malthus and the Idea of Progress." *Journal of the History of Ideas*, 28 (1966): 92–108.

Lewontin, R. C. "Adattamento." In *Encyclopedia Einuadi*, 1:198–214. Torino: Giulio Einaudi, 1977.

Limoges, Camille. "Darwin, Milne-Edwards et le Principe de Divergence." *Actes du XIIᵉ Congrès International d'Histoire des Sciences*, 8 (1968): 111–16.

"Une Lecture Nouvelle de Darwin." *Sciences*, 58–9 (1969): 70–3.

"Darwinisme et Adaptation." *Revue des Questions Scientifiques*, 141 (July 1970): 353–74.

"L'Economie Naturelle et le Principe de Corrélation chez Cuvier et Darwin." *Revue d'Histoire des Sciences et de leur Applications*, 23 (1970): 35–48.

La Sélection Naturelle: Etude sur la Première Constitution d'un Concept (1837–1859). Paris: Presses Universitaires de France, 1970.

"Introduction." In C. Linné, *L'Equilibre de la Nature*, trans. Bernard Jasmin. Paris: J. Vrin, 1972.

"Economie de la Nature et la Idéologie Juridique chez Linné." *Actes du XIIIᵉ Congrès International d'Histoire des Sciences*, Section IX, Histoire des Sciences Biologiques (1974), pp. 25–30.

"Natural Selection, Phagocytosis, and Preadaptation: Lucien Cuenot, 1886–1901." *Journal of the History of Medicine and Allied Sciences*, 31 (1976): 176–214.

Lyell, Charles. *Principles of Geology*. 3 vols. London: John Murray, 1830–3.

Principles of Geology. 5th ed. 2 vols. Philadelphia: James Kay, Jun. & Brother, 1837.

"Anniversary Address of the President." *Quarterly Journal of the Geological Society of London*, 7 (1851): xxv–lxxvi.

Geological Evidences of the Antiquity of Man. Philadelphia: George Childs, 1863.

Macculloch, John. *Proofs and Illustrations of the attibutes of God*. 3 vols. London: James Duncan, 1837.

McFarland, J. D. *Kant's Concept of Teleology*. Edinburgh: University of Edinburgh Press, 1970.

McKinney, H. Lewis. *Wallace and Natural Selection*. New Haven, Conn.: Yale University Press, 1972.

MacLeay, William Sharp. *Horae Entomologicae: Or Essays on the Annulose Animals*. 1 vol. in 2 parts. London: S. Bagster, 1819–21.

"Remarks on the Identity of Certain General Laws which Have Been Lately Observed to Regulate the Natural Distribution of Insects and Fungi." *Transactions of the Linnean Society of London*, 14 (1825): 46–68.

"A Reply to some Observations of M. Virey in the 'Bulletin des Sciences Naturelles, 1825.'" *Zoological Journal*, 4 (1828–9): 47–51

"Annulosa." In Andrew Smith, *Illustrations of the Zoology of South Africa*. London: Smith, Elder and Co., 1838.

Malthus, Thomas R. *An Essay on the Principle of Population: Or a View of its Past and Present Effects on Human Happiness.* 3rd ed. 2 vols. London: J. Johnson, 1806.

An Essay on the Principle of Population. 1st ed. (1798), ed. Philip Appleman. New York: W. W. Norton, 1976.

Mandelbaum, Maurice. *History, Man, & Reason.* Baltimore: The Johns Hopkins Press, 1971.

Manier, Edward. *The Young Darwin and His Cultural Circle.* Dordrecht, Holland: D. Reidel, 1978.

Mayr, Ernst. *Animal Species and Evolution.* Cambridge, Mass.: Harvard University Press, 1966.

Evolution and the Diversity of Life: Selected Essays. Cambridge, Mass.: Harvard University Press, 1976.

"Darwin and Natural Selection." *American Scientist,* 65 (May–June 1977): 321–7.

Milne Edwards, Henri. "Observations sur les Changemens de Forme que Divers Crustacés Eprouvent dans le Jeune Age." *Annales des Sciences Naturelles,* 2nd ser., 3 (1835): 321–34.

"Considérations sur Quelques Principes Relatifs à la Classification Naturelle des Animaux." *Annales des Sciences Naturelles,* 3rd ser., 1 (1844): 65–99.

"Observations sur la Circulation." *Annales des Sciences Naturelles,* 3rd ser., 3 (1845): 257–88.

Introduction à la Zoologie Générale. Paris: Victor Masson, 1851.

Moore, James R. *The Post-Darwinian Controversies: a Study of the Protestant Struggle to Come to Terms with Darwin in Great Britain and America, 1870–1900.* Cambridge: Cambridge University Press, 1979.

Müller, Johannes. *Elements of Physiology,* trans. William Baly. 2 vols. London: Taylor and Walton, 1838–42.

Murchison, Roderick I. *Siluria.* London: John Murray, 1854.

Needham, Joseph. "Limiting Factors in the Advancement of Science as Observed in the History of Embryology." *Yale Journal of Biology and Medicine,* 8 (1935–36): 1–18.

Ogilvie, Marilyn. "Robert Chambers and the Successive Revisions of the *Vestiges of the Natural History of Creation.*" Ph.D. dissertation, University of Oklahoma, 1973.

Oken, Lorenz. *Elements of Physiophilosophy,* trans. Alfred Tulk. London: Ray Society, 1847.

Olby, Robert C. *The Origins of Mendelism.* New York: Schocken Books, 1966.

Oppenheimer, Jane M. *Essays in the History of Embryology and Biology.* Cambridge, Mass.: The M.I.T. Press, 1967.

"An Embryological Enigma in the *Origin of Species.*" In *Forerunners of Darwin: 1745–1859,* ed. Bentley Glass, Owsei Temkin, and William L. Straus, Jr. Baltimore: The Johns Hopkins Press, 1968 (first published, 1959).

Ospovat, Dov. "Embryos, Archetypes, and Fossils: Von Baer's Embryology and British Paleontology in the Mid-Nineteenth Century." Ph.D. dissertation, Harvard University, 1974.

Bibliography

"The Influence of Karl Ernst von Baer's Embryology, 1828–1859." *Journal of the History of Biology*, 9 (1976): 1–28.

"Darwin on Huxley and Divergence." *Abstracts of Scientific Section Papers.* XVth International Congress of the History of Science (Edinburgh, 1977), p. 216.

"Lyell's Theory of Climate." *Journal of the History of Biology*, 10 (1977): 317–39.

"God and Natural Selection: The Darwinian Idea of Design." *Journal of the History of Biology* 13 (1980): 169–94.

Owen, Richard. "Remarks on the *Entozoa*." *Transactions of the Zoological Society of London*, 1 (1835): 387–94.

"Description of the Cranium of the *Toxodon platensis*." *Proceedings of the Geological Society*, 2 (1833–38): 541–2.

"On the Anatomy of the Dugong." *Proceedings of the Zoological Society of London*, Part VI, 1838, pp. 28–46.

"Fossil Mammalia." In Part I of *Zoology of the Voyage of H.M.S. Beagle*, ed. Charles Darwin. London: Smith, Elder and Co., 1840.

"Report on British Fossil Reptiles, Part II." *Report of the British Association for the Advancement of Science, 1841*, pp. 60–204.

Description of the Skeleton of an Extinct Gigantic Sloth (Mylodon robustus, Owen). London: John van Voorst, 1842.

Lectures on the Comparative Anatomy and Physiology of the Invertebrate Animals. London: Longman, Brown, Green, and Longmans, 1843.

"Teleology of the Skeleton of Fishes." *Edinburgh New Philosophical Journal*, 42 (1846): 216–27.

On the Archetype and Homologies of the Vertebrate Skeleton. London: For the author, 1848.

"Description of Teeth and Portions of Jaws of two Extinct Anthracotheroid Quadrupeds . . . with an Attempt to Develope Cuvier's Idea of the Classification of Pachyderms by the Number of Their Toes." *Quarterly Journal of the Geological Society*, 4 (1848): 103–41.

On the Nature of Limbs. London: John van Voorst, 1849.

"Address." *Report of the British Association for the Advancement of Science, 1858*, pp. xlix–cx.

On the Anatomy of Vertebrates. 3 vols. London: Longmans, Green, 1866–8.

[Owen, Richard]. "Lyell – on Life and its Successive Development." *Quarterly Review*, 89 (1851): 412–51.

"Generalizations of Comparative Anatomy." *Quarterly Review*, 93 (1853): 46–83.

"Darwin *on the Origin of Species*." *Edinburgh Review*, 111 (1860): 483–532.

Owen, Rev. Richard. *The Life of Richard Owen.* 2 vols. New York: D. Appleton, 1894.

Paley, William. *Natural Theology*, ed. Elisha Bartlett from the edition of Charles Bell and Henry Brougham. 2 vols. New York: Harper & Brothers, 1839.

Parshall, Karen H. "Varieties Are Incipient Species: Darwin's Numerical Analysis" (unpublished essay).

Bibliography

Peckham, Morse, ed. *The Origin of Species by Charles Darwin: A Variorum Text.* Philadelphia: University of Pennsylvania Press, 1959.

Peel, J. D. Y. *Herbert Spencer: The Evolution of a Sociologist.* London: Heinemann, 1971.

Peterson, Houston. *Huxley: Prophet of Science.* New York: Longmans, Green, 1932.

Pictet, F. J. *Traité de Paléontologie.* 2nd ed. Vol. 1. Paris: J. B. Baillière, 1853.

Piveteau, J. "Vertebrate Paleontology." In *History of Science in the Nineteenth Century,* ed. René Taton. New York: Basic Books, 1969.

Porter, Roy S. "The Industrial Revolution and the Rise of the Science of Geology." In *Changing Perspectives in the History of Science: Essays in Honour of Joseph Needham,* ed. Mikuláš Teich and Robert Young. London: Heinemann, 1973.

"Charles Lyell, l'Uniformitarismo e l'Atteggiamento del Secolo XIX verso la Geologia dell'Illuminismo." In *Eredità dell'Illuminismo,* ed. Antonio Santucci. Bologna: Società Editrice il Mulino, 1979.

Prichard, James C. *Researches into the Physical History of Mankind.* 3rd ed. Vol. 1. London: Sherwood, Gilbert, and Piper, 1836.

Ravetz, Jerome R. *Scientific Knowledge and Its Social Problems.* Oxford: Oxford University Press, 1971.

Redwood, John. *Reason, Ridicule and Religion: The Age of Enlightenment in England, 1660–1750.* London: Thames and Hudson, 1976.

Richards, Robert J. "Influence of Sensationalist Tradition on Early Theories of the Evolution of Behavior." *Journal of the History of Ideas,* 40 (1979): 85–105.

Roe, Shirley, "Rationalism and Embryology: Caspar Friedrich Wolff's Theory of Epigenesis." *Journal of the History of Biology,* 12 (1979): 1–43.

Roger, Jacques. *Les Sciences de la Vie dans la Pensée Française du XVIIIᵉ Siècle.* Paris: Armand Colin, 1963.

Roget, Peter Mark. *Animal and Vegetable Physiology considered with reference to Natural Theology.* 2 vols. London: William Pickering, 1834.

Rudwick, Martin J. S. "Uniformity and Progression: Reflections on the Structure of Geological Theory in the Age of Lyell." In *Perspectives in the History of Science and Technology,* ed. Duane H. D. Roller. Norman: University of Oklahoma Press, 1971.

The Meaning of Fossils. 2nd ed. New York: Science History Publications, 1976.

Ruse, Michael. "Charles Darwin and Artificial Selection." *Journal of the History of Ideas,* 36 (1975): 339–50.

The Darwinian Revolution. Chicago: University of Chicago Press, 1979.

Russell, E. S. *Form and Function.* London: John Murray, 1916.

Rylands, Peter. "On the Quinary, or Natural, System of M'Leay, Swainson, Vigors, &c." *Magazine of Natural History,* 9 (1836): 130–8, 175–82.

Schwann, Theodor. *Microscopical Researches into the Accordance in the Structure and Growth of Animals and Plants,* trans. Henry Smith. London: Sydenham Society, 1847.

Schweber, Silvan S. "The Origin of the *Origin* Revisited." *Journal of the History of Biology,* 10 (1977): 229–316.

Bibliography

"Essay Review: The Young Darwin." *Journal of the History of Biology*, 12 (1979): 175–92.

"Darwin and the Political Economists: Divergence of Character." *Journal of the History of Biology*, 13 (1980): 195–289.

Sedgwick, Adam. *Discourse on the Studies of the University of Cambridge*. 5th ed. Cambridge: John Deighton; London: John W. Parker, 1850.

Shapin, Steven, and Barry Barnes. "Darwin and Social Darwinism: Purity and History." In *Natural Order: Historical Studies of Scientific Culture*, ed. B. Barnes and S. Shapin. Beverly Hills, Calif.: Sage Publications, 1979.

Simpson, George Gaylord. "Anatomy and Morphology: Classification and Evolution: 1859 and 1959." *Proceedings of the American Philosophical Society*, 103 (1959): 286–306.

Sloan, Phillip R. "Buffon, German Biology, and the Historical Interpretation of Biological Species." *British Journal for the History of Science*, 12 (1979): 109–53.

Smith, J. Maynard. "Optimization Theory in Evolution." *Annual Review of Ecology and Systematics*, 9 (1978): 31–56.

Smith, Sydney. "The Origin of the *Origin*." *Advancement of Science*, 16 (1960): 391–401.

"The Darwin Collection at Cambridge with One Example of Its Use: Charles Darwin and *Cirripedes*." *Actes du XIᵉ Congrès International d'Histoire des Sciences* (1965) 5: 96–100.

Spencer, Herbert. "Progress: Its Law and Cause" [*Westminster Review*, 1857]. In Herbert Spencer, *Essays: Scientific, Political, and Speculative*. Vol. 1. New York: D. Appleton, 1891.

An Autobiography. 2 vols. New York: D. Appleton, 1904.

"The Filiation of Ideas." In David Duncan, *Life and Letters of Herbert Spencer*. London: Methuen, 1908.

Stauffer, Robert C. "Ecology in the Long Manuscript Version of Darwin's *Origin of Species* and Linnaeus' *Oeconomy of Nature*." *Proceedings of the American Philosophical Society*, 104 (1960): 235–41.

Stauffer, Robert C., ed. *Charles Darwin's Natural Selection*. Cambridge: Cambridge University Press, 1975.

Strickland, Hugh. "Observations upon the Affinities and Analogies of Organized Beings." *Magazine of Natural History*, new series, 4 (1840): 219–26.

Sulloway, Frank J. "Geographical Isolation in Darwin's Thinking: The Vicissitudes of a Crucial Idea." *Studies in History of Biology*, 3 (1979): 23–65.

Temkin, Owsei, "German Concepts of Ontogeny and History around 1800." *Bulletin of the History of Medicine*, 24 (1950): 227–46.

Todes, Daniel P. "V. O. Kovalevskii: The Genesis, Content, and Reception of His Paleontological Work." *Studies in History of Biology*, 2 (1978): 99–165.

Vogt, Carl. *Embryologie des Salmones*. Vol. 1 of *Histoire Naturelle des Poissons d'Eau Douce*, by Louis Agassiz. Neuchâtel, 1842.

Vorzimmer, Peter. "Darwin's Ecology and Its Influence upon His Theory." *Isis*, 56 (1965): 148–55.

"Darwin, Malthus, and the Theory of Natural Selection." *Journal of the History of Ideas*, 30 (1969): 527–42.

"Darwin's Questions about the Breeding of Animals." *Journal of the History of Biology*, 2 (1969): 269–81.

Charles Darwin: The Years of Controversy: The Origin of Species and Its Critics 1859–1882. Philadelphia: Temple University Press, 1970.

"The Darwin Reading Notebooks (1838–1860)." *Journal of the History of Biology*, 10 (1977): 107–53.

Wallace, Alfred Russel, "On the Law which Has Regulated the Introduction of New Species." *Annals and Magazine of Natural History*, 2nd ser., 16 (1855): 184–96.

"On the Tendency of Varieties to Depart Indefinitely from the Original Type." In Charles Darwin and Alfred Russell Wallace, *Evolution by Natural Selection*. Cambridge: Cambridge University Press, 1958.

Waterhouse, George R. "Descriptions of Some New Species of Exotic Insects." *Transactions of the Entomological Society of London*, 2 (1837–40): 188–96.

[Wells, Geoffrey H.] Geoffrey West. *Charles Darwin: A Portrait*. New Haven, Conn.: Yale University Press, 1938.

Whewell, William. *Astronomy and General Physics considered with reference to Natural Theology*. London: William Pickering, 1833.

A History of the Inductive Sciences. 3 vols. London: John W. Parker, 1837.

[Whewell, William]. *Of the Plurality of Worlds: An Essay*. 5th ed. London: John Parker, 1859 (first published, 1853).

Willey, Basil. *The Eighteenth Century Background*. Boston: Beacon Press, 1961 (first published, 1940).

Wilson, Leonard G., ed. *Sir Charles Lyell's Scientific Journals on the Species Question*. New Haven, Conn.: Yale University Press, 1970.

Winsor, Mary P. "Barnacle Larvae in the Nineteenth Century: A Case Study in Taxonomic Theory." *Journal of the History of Medicine and Allied Sciences*, 24 (1969): 294–309.

Starfish, Jellyfish, and the Order of Life: Issues in Nineteenth-Century Science. New Haven, Conn.: Yale University Press, 1976.

"Louis Agassiz and the Species Question." *Studies in History of Biology*, 3 (1979): 89–117.

Young, Robert. "Malthus and the Evolutionists: The Common Context of Biological and Social Theory." *Past and Present*, no. 43 (May 1969), pp. 109–45.

"Darwin's Metaphor: Does Nature Select?" *The Monist*, 55 (1971): 442–503.

"Darwinism and the Division of Labour." *The Listener*, 88 (August 17, 1972): 202–5.

"The Historiographic and Ideological Contexts of the Nineteenth-Century Debate on Man's Place in Nature." In *Changing Perspectives in the History of Science: Essays in Honour of Joseph Needham*, ed. Mikuláš Teich and R. Young. London: Heinemann, 1973.

Index

The index was prepared by Elin L. Wolfe, with the assistance of Shirley A. Roe
and Frank J. Sulloway.

Index

Darwin, Charles (*cont.*)

theory in, 73–83, 234; treatment of embryology in, 153, 155, 156, 162

"General Observations," 24

Hooker's visits to, 93–4

Journal, 170

Journal of Researches, 92

Malthus's impact on, 61–73

Natural Selection, 2, 88, 89, 91, 95, 170, 206, 221, 224, 273 (n 65); comparison with "Essay of 1844," 197, 207–9; discussion of variation in, 200–5; principle of divergence in, 171, 177, 179, 180, 181, 184–9, 192; views on nature in, 198–9

notebooks on transmutation, ix, 1, 2, 4, 6, 23, 25–6, 27, 30, 31, 33, 37, 38, 40, 44, 54, 67, 70–1, 77, 84, 88, 89, 91, 92, 98, 108, 111, 113, 151, 152, 158, 191, 204, 226; B (first) notebook, 26–7, 28–9, 30, 33, 39, 40–1, 41–2, 43, 44, 45, 46, 48, 51, 55, 56, 61–2, 74–5, 79, 83, 99, 108, 109–10, 110–11, 147, 150, 151, 152, 178, 203, 211, 212, 226, 231, 242 (n 70), 244 (n 23), 254 (n 76, n 86), 255 (n 89); C (second) notebook, 29, 30, 33, 41, 42, 43, 45, 46, 47, 48, 51–2, 56–7, 67, 87, 111, 112, 147, 152–3; D (third) notebook, 32, 33, 42, 44, 46, 47–8, 49, 53, 60, 61–2, 67, 87, 109, 111, 152, 211–12, 213–14; E (fourth) notebook, 31, 33, 68, 69–70, 71, 75, 83, 147, 152, 195–6, 214, 248 (n 25)

notes, unpublished, ix, 159, 272 (n 55); envelopes of, 88–9, 94, 108, 170, 171, 251 (n 4)

Origin of Species, ix, 1, 2, 4, 6, 7, 27, 37, 62, 67, 73, 88, 96, 140, 151, 153, 163, 169, 189, 207, 209, 221–2, 228, 232, 234; comparison with "Essay of 1844," 75, 80, 83–6; Cuvier's principle in, 149; Darwin's embryological principles in, 156–7, 165; dis-

cussion of variation in, 191, 204; Part I and Part II of species work in, 88, 98, 99, 100, 167, 168; theological assumptions in, 72, 224, 226; three-element hypothesis in, 112–13; views on nature in, 223, 224

Owen's relations with, 92–4, 261 (n 23)

Part I of species work, 87–8, 91, 185, 251 (n 2)

Part II of species work, 87–114, 117, 145, 165, 167, 168, 169, 170, 175, 189, 190, 228, 230

pigeon project of, 156–7

principles of, *see under* divergence; hereditary fixing; variation

professionalism of, 4, 91

"Sketch of 1842," 1, 2, 72, 73, 78, 82, 84, 87, 88, 90, 98, 153, 156, 162, 201, 223

theological assumptions of, 32–3, 72–3, 150, 207, 224–6, 273 (n 64); *see also* views on nature

theories of, *see under* descent; evolution; generation; natural selection; progress; transmutation; variation

The Variation of Animals and Plants under Domestication, 89, 203

views on nature, 3, 37–9, 59, 60–1, 275–6 (n 49); changes in views, 23–33, 83–4

work on large, aberrant, and small genera, 92, 177–86, 194, 202, 208, 265–6 (n 27), 266 (n 30, n 34), 266–7 (n 39)

Zoology of the Voyage of H.M.S. Beagle, 134

Darwin, Erasmus, 93

Darwin, Francis

"Reminiscences," 95

Darwin and his contemporaries

correspondence of views, 30, 54, 58–9, 67, 68, 85, 183, 196, 199, 200, 225

divergence of views, 73

influence of thought and development of ideas, 1–5, 113–14, 169,

LaVergne, TN USA
25 October 2009
161971LV00014B/11/A